Fixed Asset Valuation

The valuation of Fixed Assets is a core discipline at the intersection of finance, accounting and asset management. As industries face increasingly complex regulatory frameworks and global market dynamics, developing a solid understanding of valuation principles is essential for both academic study and professional practice. This comprehensive book offers a structured, academically rigorous introduction to the appraisal of property, plant and equipment, bridging theoretical depth with real-world application.

Drawing on a proven five-phase project methodology refined through decades of international consulting experience, the authors provide a systematic framework for approaching valuation challenges. The text combines international standards, theoretical foundations and applied case studies, making it a robust resource for students, researchers and early-career professionals seeking to build advanced expertise in asset valuation.

This guide is ideally suited for graduate and postgraduate students in corporate finance, accounting, real assets, insurance, taxation and business management. It is also a valuable reference for academics and practitioners who wish to deepen their understanding of valuation methodology and its role in decision-making.

The readers of this book will:

- Gain a thorough understanding of international valuation standards and their practical implications.
- Learn a structured five-phase approach to conducting asset valuations from project preparation to reporting.
- Analyze valuation methodologies for property, plant and equipment within various regulatory and market contexts.
- Develop the ability to critically assess valuation reports and apply theoretical principles to complex real-world scenarios.
- Understand how valuation integrates with financial reporting, taxation, insurance and investment decision-making.
- Benefit from case studies and practical examples drawn from global consulting and academic experience.

Florian Hackelberg is a Professor of Real Estate Valuation and Investment at the Berlin School of Economics and Law. With over 20 years of international teaching, research and consulting experience, he specializes in real estate and fixed asset valuation. He holds a PhD from the Technical University of Berlin and a master's degree from the University of Cambridge.

Dirk Kadel is a partner at PwC Germany, where he leads the firm's fixed asset valuation practice. He has over 25 years of experience in global consulting and valuation projects, bringing deep industry expertise and technical insight. He holds a master's in engineering and is a Member of the Royal Institution of Chartered Surveyors (MRICS).

Fixed Asset Valuation

A Practical Guide to the Appraisal of Property, Plant and Equipment

Florian Hackelberg and Dirk Kadel

Routledge
Taylor & Francis Group
LONDON AND NEW YORK

Designed cover image: © iStock

First published 2026
by Routledge
4 Park Square, Milton Park, Abingdon, Oxon OX14 4RN

and by Routledge
605 Third Avenue, New York, NY 10158

Routledge is an imprint of the Taylor & Francis Group, an informa business

British Library Cataloguing-in-Publication Data
A catalogue record for this book is available from the British Library

ISBN: 9781041017431 (hbk)
ISBN: 9781041031314 (pbk)
ISBN: 9781003622413 (ebk)

DOI: 10.1201/9781003622413

Typeset in Optima
by Newgen Publishing UK

Contents

Figures

About the Authors

Florian Hackelberg is a Professor of Real Estate Valuation and Investment at the Berlin School of Economics and Law. He previously worked as a Senior Manager in the Valuation Advisory division at PricewaterhouseCoopers GmbH Wirtschaftsprüfungsgesellschaft (PwC) in Berlin and Shanghai.

After studying civil engineering and construction management in Germany, England and China, Florian Hackelberg completed his PhD at the Technical University of Berlin. He also holds a master's degree in international relations from the University of Cambridge.

Florian Hackelberg is a professional member of the Royal Institution of Chartered Surveyors (MRICS), a member of the RICS Professional Group Valuation, a member of the Society of Property Researchers, Germany, and an honorary valuer for the Hoexter Valuation Committee for Property Values. He is the editor and author of numerous books and articles.

Dirk Kadel is a partner with the Valuation Advisory division at PricewaterhouseCoopers GmbH Wirtschaftsprüfungsgesellschaft in Germany (PwC) and leads the firm's tangible assets valuation practice. He has more than 25 years with PwC and has worked across the globe on international projects.

Dirk Kadel has extensive experience in leading valuation projects of Real Estate, infrastructure as well as machinery and equipment and has been responsible for a countless number of national and international valuation projects. In particular, his focus is on Fair Value valuation for financial reporting, tax and financing purposes. He lectures on the topic both internally at PwC as well as universities and institutions.

Dirk Kadel is a professional member of the Royal Institution of Chartered Surveyors (MRICS) and a registered architect. He graduated with an engineering degree from the Technical University of Darmstadt.

Foreword

Assessing the value of Fixed Assets such as machinery and equipment is an essential part of the economic and business environment. These valuations influence financial reporting, taxation, insurance and strategic business decisions. While literature on Real Estate valuation abounds, there is a clear lack of comprehensive resources tailored to the unique demands of machinery and equipment valuation. This book fills that need, offering in-depth guidance that blends foundational knowledge with actionable techniques.

Fixed assets play a crucial role in business operations, and their valuation requires a careful mix of theoretical understanding and practical application. The text introduces a systematic method for organizing and executing valuation assignments, addressing the particular challenges of evaluating a wide range of Fixed Assets.

The increasing internationalization of asset portfolios further underscores the need for a comprehensive understanding of valuation practices, supported by industry standards from organizations like the Royal Institution of Chartered Surveyors (RICS).

By weaving together core principles and effective practices, this book is an indispensable guide for valuation professionals. It walks readers through each stage of the process – from preparation and data collection to the application of various valuation approaches and the final reporting phase. This book also addresses the review of third-party valuations, offering procedures and checklists to ensure proper execution and market-based values.

Based on the authors' 25 years of experience as practitioners and teachers, this book stands out as a vital reference for anyone engaged in the valuation of Fixed Assets. It enhances the reader's comprehension of valuation and its importance in asset management, making it a valuable tool for both academics and practitioners.

Susanne Eickermann-Riepe
Senior Vice President at RICS

Preface

Dear Reader,
You are about to step into the world of Fixed Assets – assets that shape industries, guide strategic investments and underpin crucial business decisions. Whether you're a student curious about the field or a seasoned professional seeking deeper understanding, this book was written with you in mind. Fixed assets, ranging from sophisticated machinery to sprawling facilities, are more than numbers on a balance sheet; they are the backbone of entire enterprises. Yet, understanding how to value these assets accurately is often a daunting challenge, especially when portfolios span countries and regulatory environments.

You may have noticed that while Real Estate valuation is supported by abundant literature, resources dedicated to machinery and equipment remain surprisingly scarce. If you've ever felt uncertain about how to approach fixed asset valuation, or wished for a guide that demystifies the complexities, you are not alone. That's why we set out to create this book: to provide you with a comprehensive, practical framework tailored specifically to the diverse world of property, plant and equipment.

Inside, you'll find clear explanations of the theoretical principles and proven methodologies that underpin fixed asset valuation. We've distilled over 20 years of experience working alongside clients across continents and industries into actionable insights you can rely on. Our goal is to equip you with the tools and knowledge you need to tackle valuation projects with confidence, whether your focus is financial reporting, tax compliance, investment decisions or regulatory requirements.

Drawing on over two decades of experience working for international clients from local businesses to leading multinational corporates, we have had the privilege of overseeing extensive valuation projects across all continents and numerous industries. During this process, we have provided support in a diverse variety of valuation challenges. Despite the uniqueness of each valuation project, there are international valuation standards and proven methodological approaches which provide a reliable framework. The purpose of this book is to provide a clear and comprehensive explanation of the theoretical core principles, and to demonstrate their application in valuation practice.

We are delighted to share the insights and lessons distilled from our hands-on experience with you and we hope this book will serve you as a valuable resource engaging with fixed asset valuation.

Florian Hackelberg, Berlin
Dirk Kadel, Frankfurt

1 Introduction

Fixed Assets are critical components of economies, businesses and individuals' wealth. These assets encompass a wide range of items, including land, Real Estate, infrastructure, factories, machinery, equipment, vehicles, tools and inventory. Often referred to as tangible assets, Fixed Assets are also commonly known as property, plant and equipment. This category can be broadly further divided into Real Estate and Machinery and Equipment (M&E), with infrastructure also considered a part of Fixed Assets.

Understanding the value of an asset and conducting its valuation is essential for various reasons and purposes. Those valuation reasons and related Fixed Assets are often not limited to national borders, but many Fixed Asset valuation projects have an international component as well. With increasing internationalization, there has been a significant rise in the need to assess the value of international investment portfolios. This is particularly crucial in cross-border transactions, investments and joint ventures, where determining the value of acquired Fixed Assets is paramount. Moreover, the valuation of Fixed Assets has gained prominence due to the comprehensive international financial reporting practices developed over recent decades.

While there is a variety of laws, guidelines and literature available for Real Estate valuation, the same cannot be said for the valuation of M&E. Although various local guidelines are published by regulatory bodies and some international standards exist for the valuation of Fixed Assets, these resources often approach the topic from a theoretical and definitional standpoint. Industry standards such as those provided by International Valuation Standards Council (IVSC), American Society of Appraisers (ASA), Royal Institution of Chartered Surveyors (RICS) and The European Group of Valuers' Associations (TEGOVA) are crucial for establishing a shared understanding among international market participants and stakeholders. However, these guidelines frequently emphasize how valuations theoretically should be conducted rather than how this theory is applied in practice, focusing less on the practicalities of the valuation process itself.

Although valuation theory for Fixed Assets is less financially complex compared to business enterprise valuation or the valuation of financial products, the practical execution is often complex as well. It is not uncommon that a valuation project can involve tens of thousands of individual assets leading to significant practical challenges in the valuation process.

In light of these considerations, this book not only presents the theoretical valuation principles and definitions but also emphasizes practical strategies for conducting market-based valuations of Fixed Assets effectively and efficiently. Hence alongside theoretical insights, this book offers a variety of practical tips and advice to aid practitioners in the field.

DOI: 10.1201/9781003622413-1

In order to ensure such a reliable assessment, a structured process must be followed in practice. In practice, a five phase approach has proven particularly effective for managing complex valuation projects in a structured and efficient manner (see Figure 1.1).

1.1 The Five Phases Approach

The valuation of Fixed Assets may be necessary for a variety of reasons. In practice, valuation projects which involve the valuation of several fixed asset classes such as Real Estate and M&E are frequently essential for financial reporting purposes, necessitating adherence to internationally accepted financial reporting standards, such as in Purchase Price Allocations (PPA) in accordance with the International Financial Reporting Standards (IFRS). These valuations, alongside considerations related to transactions and tax and insurance law, constitute a significant portion of valuation projects performed by many appraisers. As such, the strategies discussed in this book are aligned with procedures for valuing tangible assets for such purposes. Even though the fundamental approaches and methods remain consistent, certain nuances must be addressed.

This approach is beneficial not only for implementing Fixed Asset valuation projects successfully but also for engaging all stakeholders involved in the process early on. These stakeholders frequently include but are not limited to: the client, the target management, auditors and involved authorities and appraisers for other valuation subjects in the broader projects, such as intangible assets or business enterprise valuation. Early involvement is especially crucial in complex international projects to ensure all aspects are adequately considered and agreed by all parties so that the valuation project can be completed as scheduled.

The structure of this book aligns with this five phase approach. Since valuing Fixed Assets is often part of a larger, multifaceted project with other related work streams, including e.g. the valuation of intangible assets, tax assessments, environmental burdens analyses, analyses of pensions on an international level, meticulous planning during the

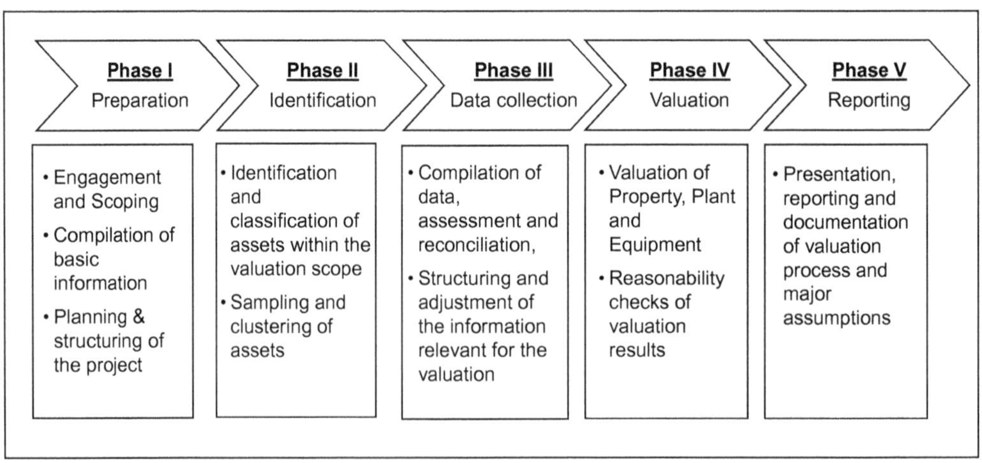

Figure 1.1 Five Phases Approach to Fixed Asset Valuation.
Source: Adapted from Plock et al. [2025], p. 20.

preparation phase is vital. This involves precisely defining the valuation object and structuring the valuation procedure based on the valuation context.

1.2 Structure of the Book

Unlike the comparatively straightforward valuation of e.g. an apartment or single-family home, assessing an industrial production site with its diverse components such as land, buildings, production lines, machinery, vehicles and tools requires a more nuanced approach. Given resource constraints, valuing each asset in detail is often impractical due to the time and expense involved. Therefore, for smaller-scale M&E assets, assumptions and potential necessary simplifications are regularly employed to achieve reliable valuation results on the one hand but with reasonable efforts and within a reasonable timeframe on the other.

The value of Fixed Assets can vary significantly, depending on their size and scope, ranging from billions for large infrastructure or industrial facilities to individual low-value items. Besides valuation framework and background, specific value definitions as well as valuation concepts and scenarios are detailed in Chapter 2, "Basis of Value."

The work of a valuation starts already before the actual valuation process, with proper planning and structuring of the project, including the engagement and compilation of basic data. Those preparation steps are outlined in Chapter 3, "Preparation," of this book.

The focus of a valuation exercise typically lies on large, significant assets or, in the context of financial reporting valuation e.g. a PPA, those assets which are expected to undergo substantial Fair Value Adjustments (the difference between the Net Book Value (NBV) and respective Fair Value of an asset). This process, detailed in Chapter 4 "Identification," is part of the identification phase, utilizing portfolio analysis and valuation techniques. By the end of this phase, it's clear which assets require a detailed valuation (Cluster A), which can be valued on high level basis only e.g. through desktop methods (Cluster B) and which less-significant assets (Cluster C) can be statistically analyzed based on valuation results from Clusters A and B. Depending on the valuation's scope, clustering may refer to the country level, location level or business unit level.

In practice, a major challenge is not only collecting necessary information but also verifying and reconciling it with various data sources. Particularly, reconciling technical data with financial accounting records and the related market information can be complex and time-consuming, yet it's crucial for ensuring a solid valuation foundation. This process is described in Chapter 5, "Data Collection, Validation and Reconciliation." The Fixed Asset Register (FAR) serves as the primary basis for fixed asset valuation, listing essential information such as purchase date, acquisition costs or Gross Book Value (GBV) and depreciated NBV. Site visits help verify this information and gather additional relevant data, such as the extent and type of structural use and building condition for Real Estate or the actual economic use and obsolescence for machinery.

As described in Chapter 6, "Valuation," Phase IV of the process involves selecting the appropriate valuation method based on the valuation context, the Fixed Assets to be valued and available data, whether using market approach, income approach or cost approach. Different methods may apply to various types of Fixed Assets within a location or production site; e.g. the sales comparison method for land, the investment method for buildings and the cost approach for M&E. Valuation models are crafted based on received client data and researched market information, often using individual financial mathematical models rather than valuation software due to the project-specific nature of the data and structure.

The final Phase V involves creating audit-proof and comprehensible documentation of the valuation procedure and results, aligned with the valuation's purpose. While e.g. an indicative "Transaction valuation" may only require a brief memo, a full-scope financial reporting valuation or valuations for tax or financing purposes necessitate a detailed expert opinion including proper documentation of applied approaches, methods and assumptions.

As the work of a valuer does not only include valuations but also the review of third-party valuations as well as valuation advisory and consulting, the last Chapter 8 of this book "Third-Party Valuation Review" offers various review procedures and a detailed checklist for valuers to review a third-party valuation including valuers' formal qualifications, valuation procedures and approaches and key approaches, ensuring proper execution and market-based values.

2 Basis of Value

Before exploring the various phases of the valuation process in subsequent chapters, this section will elaborate on the foundational principles that govern the valuation of Fixed Assets. Recognizing the significance of these principles and their considerations is essential for a robust understanding of valuation practices.

Hence this chapter will provide an overview of Fixed Assets, which are the primary focus of valuation efforts. Special attention will be directed toward the concept of the purpose of valuation concepts and procedures, which explains the rationale behind different valuations, and the various definitions of value that are pertinent in this context. Furthermore, this chapter will outline the valuation frameworks and methodologies applied across different situations.

As iterated by The European Group of Valuers' Associations (TEGOVA), a basis of valuation is defined as "a statement of the fundamental assumptions for undertaking a valuation for a defined purpose."[1] This definition underscores the necessity for clarity and specificity within the valuation process, ensuring that each valuation is contextualized appropriately.

The discussions and definitions presented herein will form the conceptual foundation for the subsequent chapters, equipping readers with essential knowledge and terminology that underlie the valuation of Fixed Assets including both Real Estate as well as Machinery and Equipment (M&E). By establishing this framework, we aim to enhance the reader's comprehension of the complexities associated with fixed asset valuation and its importance in the broader scope of asset management.

The general assumption underlying all the valuation concepts provided hereafter on the subject of valuation is that it is a valuation performed in a largely free market environment, where supply and demand meet and thus fundamentally determine what happens in the market. This does not mean that the market is not subject to local regulations which may need to be taken into account in the valuation, and which may also affect the derived value. However, the basic assumption in all of the following discussion of valuation is that the prices of economic goods are not set by regulators, such as government agencies, but are determined in the marketplace by the mechanism of supply and demand.

2.1 What Is Valuation?

Valuation refers to the process of determining a reliable value for any economic asset. The outcome of this valuation process is typically a comprehensible and credible estimate of the Market Value of an asset, founded on systematic analysis and substantiated methodologies.

DOI: 10.1201/9781003622413-2

Valuation
Valuation = Information + Analytics + Experiences (+ Technology)

Figure 2.1 Valuation.

In conducting a valuation, the objective is to arrive at an as-objective-as-possible assessment of the value of the asset in question. This is typically achieved through the application of standardized procedures and methodologies that have been developed within the field of valuation. These processes and methodologies are designed to minimize subjective bias and to ensure that the valuation reflects the genuine conditions of the market. In short: "Valuation is a structured mathematical approach to estimate the potential price of an asset."[2]

All valuation approaches and methodologies somewhat aim to compare the subject asset to be valued with reference points from its specific market. This basic idea is not limited to Fixed Assets but more or less for the valuation of all assets or economic goods. While it is comparatively easy for goods and commodities like natural resources or shares at a stock market which are sold and purchased almost any time leading to transparent pricing which serves as such market reference point, this is different for Fixed Assets. Since certain Fixed Assets, especially Real Estate, is by definition unique because of its unique location, it makes the comparison with other assets particularly difficult if not impossible.[3] This fundamental problem of the asset class leads to the core challenges in Fixed Asset valuation.[4]

As a resulting consequence, valuation is a complex process in which not only information and its analysis but also the expertise and experience of the valuer play an essential role to derive the value of an asset (Figure 2.1).

2.1.1 Information: The Foundation of Valuation

Information forms the backbone of any valuation exercise. In the area of fixed asset valuation, two primary types of information are paramount:

Asset or Object Information: This involves detailed data about the specific asset being valued. It includes location, physical characteristics, historical acquisition costs, current condition, functional capabilities and any relevant legal or regulatory considerations. Asset information ensures that the valuation is tailored to the unique attributes of the asset, accounting for factors such as age, depreciation and any enhancements or modifications.

Market Information: This encompasses data regarding respective current market conditions, including trends in asset prices, demand and supply dynamics and economic indicators. Market information provides the contextual backdrop against which the value of a fixed asset is assessed. Among others, it includes comparable sales data, market forecasts and industry-specific insights that help valuers understand the broader environment influencing asset values.

2.1.2 Analytics: Transforming Data into Insight

Analytics can be described as the process of transforming raw data into actionable insights. In fixed asset valuation, analytics involves a series of steps:

Data Structuring: This step involves organizing the data in a coherent manner, allowing for efficient analysis and interpretation. Proper structuring is crucial for aligning data with the chosen valuation methodology, whether it be cost, income or market approach.

Data Analysis: This entails examining the gathered information to identify patterns, trends and anomalies. It involves statistical and quantitative techniques to interpret the data accurately, ensuring a comprehensive understanding of both market and asset-specific factors.

Valuation Assessment: Utilizing the structured data, valuers apply valuation models and methodologies to determine the asset's value. This involves calculating depreciation, applying market comparables or projecting future income streams, depending on the asset type and available information.

2.1.3 *Experience: The Intangible Edge*

Experience is the intangible yet invaluable component of the valuation equation.[5] It encompasses two critical dimensions:

Market Experience: This refers to the valuer's familiarity with and understanding of market dynamics within the specific industry or asset class. It includes insights gained from previous valuations, awareness of historical trends and the ability to anticipate future market movements.

Valuation Experience: This involves the valuer's expertise in applying valuation principles, methodologies and judgment. It includes the ability to navigate complex valuation scenarios, recognize potential risks and uncertainties and adapt methodologies to suit the unique characteristics of each asset.

2.1.4 *Technology: The Future of Valuation*

Despite a somewhat tentative initiation, digitalization is progressively permeating the Real Estate sector.[6] This has resulted in the emergence of a multitude of novel business models, particularly within the domain of property valuation. The advent of sophisticated data analysis tools has rendered it possible for non-specialist users to undertake property valuation tasks independently via online portals. These tools have the potential to facilitate the automated assessment of property value in the future, with this assessment being based on specific characteristics of the property's location, use and standard of furnishings. The application of Automated Valuation Models (AVMs) in this context facilitates the execution of valuation tasks not only by non-specialist private individuals but also by professionals within the property valuation sector.[7] As a case in point, AVMs are utilized for the purpose of determining the value of transactions or mortgages.[8]

The impact of technology and big data on valuation in practice can be categorized into two distinct effects. Firstly, the process of optimizing traditional valuation processes is being driven forward by data-supported tools and models. Secondly, the entire valuation process is being fully automated by automated valuation machines and artificial intelligence.

Optimizing the Traditional Process: Within the "traditional valuation process," there are numerous opportunities to use digital tools and thus utilize technological developments to support the valuation process and make it more efficient. In the contemporary business environment, there is an increasing emphasis on the streamlining and automation of work processes. Consequently, the professional qualifications of experts see an elevated demand for expertise in Information Technology (IT).[9] The utilization of computer-aided valuation programs has become a prevalent practice in contemporary business operations, manifesting in various forms, including direct, system-controlled links with clients.

Disrupting the Traditional Valuation: In some areas, the traditional valuation process is already being disrupted by automated valuation (AVMs). However, it should be noted that these do not pertain to M&E, but rather to Real Estate, and more specifically to highly standardized forms of property, provided that there is sufficient market data transparency, for instance in the case of flats or detached houses. Despite the utilization of these automated valuations by banking institutions to ascertain the Mortgage Lending Value for modest loan amounts, the intricacy and inherent property characteristics of the tangible assets previously mentioned have not yet facilitated the widespread adoption of these AVMs. Apart from current technical insufficiency the question of who takes responsibility and subsequently liability in the valuation using AVMs is a salient one as well, and as such, their utilization across the board in practice is not yet possible. TEGOVA therefore emphasizes: It is evident that AVMs are not capable of being utilized in order to produce a valuation report that complies with EVS in an independent manner, without a valuation process that is founded, inter alia, on the valuer's inspection of the property and the application of his/her professional judgment. In instances where it is employed, an AVM should be regarded as a mere instrument that can assist in the estimation of value, for which the valuer must assume responsibility.[10]

The task of the valuer is to simulate the market and to depict this simulation. And as long as people make decisions, people will continue to value them.[11] Thus the focus of this book is on the traditional valuation process, in which technological developments, particularly in the area of data processing, are taken into account.

The described equation encapsulates the comprehensive nature of fixed asset valuation. Information provides the necessary data foundation, analytics transform this data into meaningful insights and experience infuses the process with expertise and judgment. Together, these components ensure that valuations are accurate, reliable and reflective of both current market conditions and the specific attributes of the asset. By understanding and effectively integrating these elements, valuers can deliver robust valuations that support informed decision-making and strategic asset management.

Ultimately, the valuation process serves not only to ascertain a specific value for the asset but also to provide stakeholders with reliable information that can inform decision-making processes, investment strategies and financial reporting. By engaging in a rigorous and methodical valuation process, one ensures the transparency and credibility of the value assigned to economic goods, thereby facilitating trust among various parties involved in financial transactions.[12]

2.1.5 Valuation of Fixed Assets

Fixed Assets valuation centers on assessing individual assets or a portfolio of assets. These assets can be valued independently or in relation to their interdependence, such as within a production process. While Real Estate assets are typically valued individually, M&E Assets are often part of a larger industrial or production process and hence considers their interdependence (also see Going Concern Principle)

The valuation of Fixed Assets differs significantly from business valuation. It focuses on individual assets such as land, building and improvements, M&E, determining their economic worth based on factors like age, condition, market demand and potential future earnings.

In contrast, business valuation entails a broader analysis of the entire economic unit. This approach values the company as an integrated entity, considering the interplay between its assets and liabilities to derive a comprehensive understanding of its overall worth, including how Fixed Assets contribute to revenue generation and operational efficiency.

The distinction between these two valuation types is particularly important for companies with a large stock of Fixed Assets, which thus significantly impact the overall valuation. Despite some overlap, each type employs different methodologies tailored to its unique focus.

2.1.6 Valuation Is Not an Exact Science

Unlike in physics or mathematics, where precise calculations yield exact results, valuation does not represent a single correct outcome, rather, there exists a spectrum of reasonable valuation outcomes.[13]

The reason for this lies in both the applied valuation methodology as well as the underlying valuation assumptions and scenarios. While in theory various valuation techniques and methodologies should converge on the same value, in practice, they often diverge. The reason for this is that different valuation concepts are based on different market comparisons or market reference points. Additionally, numerous assumptions must be made based on projections and forecasts regarding future market developments. Even with a thorough analysis of the assets and comprehensive research into the relevant market, these remain assumptions. Each market participant may have a slightly different perspective on market trends, which can result in varying assumptions.[14]

Furthermore, the valuation process invariably involves subjective elements that cannot be eliminated, even with the best intentions and knowledge of the valuer. This subjectivity is both inevitable and desirable, as the assumptions underlying the valuation also reflect the valuer's experiences and professional judgment concerning the respective markets.

As previously discussed, while the goal of valuation is to objectively and methodically determine a value, the process is ultimately carried out by individuals who are never fully free from subjective influences and external factors. Consequently, these elements will invariably affect the valuation outcome. In both literature and legal contexts, it is commonly accepted that a value deviation of +/- 20% is still considered reasonable within market standards.[15]

To mitigate the risk of perceived bias,[16] valuers should strive to derive all assumptions as transparently as possible during the valuation process and substantiate their applied assumptions quantitatively. This diligence enhances the credibility and reliability of the valuation, grounding it in objective analysis while acknowledging the inherent uncertainties involved.

2.1.7 Similarities and Differences in Real Estate and M&E Valuations

Valuation of Real Estate and M&E, while sharing several foundational similarities, diverges significantly due to the inherent characteristics of the assets (see Figure 2.2). Both domains employ the fundamental valuation methods: the market (comparative), income (investment) and cost (depreciated cost new) approaches. These methodologies provide a structured framework for assessing the respective asset's value, irrespective of the asset class. International standard-setting bodies such as the International Valuation Standards Council (IVSC), TEGOVA and the International Financial Reporting Standards (IFRS) provide guidelines that harmonize valuation practices across these sectors.

The underlying reasons for conducting valuations, such as financial reporting, investment analysis and transaction purposes, remain consistent, and the definitions and concepts of value are uniformly defined, ensuring clarity and consistency in valuation discourse.

However, the distinct nature of M&E compared to Real Estate introduces significant differences in the valuation process. The inherent mobility of M&E contrasts sharply with the immobility of Real Estate, impacting valuation methodologies and considerations. In M&E valuation, the Market Approach and especially Cost Approach predominate,

Similarities	Differences
1. Valuation methods (market, income and cost approach = comparative, investment and depreciated cost new method) are basically the same 2. Well-known standard setters (IVSC, TEGoVA, IFRS) and valuation organizations (e.g. RICS, ASA) define standards and certify experts 3. The reasons for valuation and the procedures are often similar 4. Definitions and concepts of value are identical in formulation and content	1. Valuation in machinery and equipment is dominated by the market approach and cost approach; the income approach is used very rarely 2. A quantity structure from the accounting (e.g. Fixed Assets Register) is used much more frequently for machinery and equipment valuation. Hence the valuation is somewhat closer linked to the accounting 3. Valuation objects must be considered in their environment (e.g. certain necessary fixtures might be 'lost values' upon sale) 4. Control technology and networking of a plant are constantly increasing ('The control of a machine is the "technical building equipment" of a building') 5. The third-party use of the assets with regard to different use is rather limited 6. Machines are basically 'mobile'

Figure 2.2 Real Estate Valuation vs. Machinery and Equipment Valuation.
Source: Authors, adapted from Plock et al. [2025], p. 7.
Note: Cf. Schulte et al. [2016], p. 5.

while the Income Approach is seldom utilized. This reflects the tangible, depreciable nature of M&E assets, which are often valued based on their current utility and replacement cost rather than potential income generation. The valuation of M&E frequently relies on detailed quantity structures derived from accounting records, linking valuation closely with financial accounting practices. In contrast, Real Estate valuation is more attuned to market dynamics and locational factors.

Further, M&E must be valued within their operational environment, considering necessary fixtures and potential "lost values"[17] upon sale. The increasing integration of control technology and networking within machinery, akin to technical building equipment in Real Estate, adds layers of complexity to valuation, demanding specialized knowledge of technological advancements. The adaptability of machinery for third-party use is generally limited, unlike Real Estate which often accommodates diverse uses.

Additionally, machinery markets are globally dispersed and vary in developmental stages, whereas Real Estate valuation is heavily influenced by local market conditions. Thus, while the overarching principles and frameworks of valuation may be shared, the practical applications diverge significantly between Real Estate and M&E, reflecting the unique attributes and market dynamics of each asset class.

2.1.8 Valuation Practice

The activities related to valuation can be broadly divided into three distinct areas of a valuer's work[18]:

- Valuation
- Valuation Review and
- Valuation Consulting/Advisory.

Valuation pertains to the actual determination of an asset's worth and represents the fundamental responsibility of the valuer. The objective of the valuation process is to establish the value of a specific asset. The methodologies employed, the assumptions made and the resultant findings are typically compiled in a valuation report. These procedures are fundamentally aligned with the approaches described within this book.

While the actual valuation is the major or classical work of a valuer, the valuer will not always be engaged to prepare a valuation on his own, but also for reviewing other valuers work to form an understanding of the reasonableness of the value derived.[19] This process is called Valuation Review and depending on the valuers' job environment (e.g. Accounting and Financial Reporting) can heavily contribute to the valuers' daily work. Valuation Review involves the examination or analysis of an appraisal prepared by a third party. In this context, the valuer does not create an independent valuation but rather critically analyzes the assessment conducted by another valuer. Various reasons may necessitate the review of an appraisal, outlined in greater detail in Chapter 6 of this book. Such reviews often arise within the framework of financial reporting or when preparing counter-appraisals, particularly in the context of transactions or legal dispute involving the valuation of Fixed Assets.

The third area, Valuation Consulting (or Valuation Advisory), encompasses advisory services that revolve around the domain of valuation and/or areas where the valuer can advise the client on fixed asset related matters applying his/her market and technical knowledge from his/her valuation work. These services can take many forms and may include a range of topics related to asset valuation.

Valuers specializing in Fixed Assets possess extensive knowledge in Real Estate economics, technical processes and operational practices, enabling them to address a wide array of valuation-related inquiries. These inquiries may range from classical market analyses and economic feasibility studies to strategic considerations regarding the management of Real Estate portfolios or technical due diligence in relation to M&E.[20]

It is essential to clarify, at the outset of any engagement, whether the valuer is being engaged by the client as an independent valuer or appraiser (Valuation) or as a consultant (Valuation Consulting). When operating as a valuer, they must maintain impartiality and determine an objective value to the best of their knowledge and belief.

When a valuer is engaged as a consultant rather than as an appraiser, their role may shift significantly. This engagement can, at times, carry a more subjective nature, making it imperative that the valuer clearly delineates their function in any reporting, thereby allowing clients to understand the capacity in which they are operating.

Due to their diverse academic backgrounds in economic, technical and legal disciplines, as well as their broad spectrum of expertise and experience across various fields, valuers are employed by a wide range of industries and employers. This includes traditional appraisal firms, consulting companies, brokerage firms, banks, professional services firms, insurance agencies and corporate entities.

2.2 Definition of Valuation Object

The first step in any valuation process is to gain a precise understanding and clear delineation of the asset being valued. While this may seem trivial when dealing with Fixed Assets, given that they are tangible items, it remains a critical and often nuanced task. Although defining and identifying Fixed Assets may seem less complex than the identification and definition assessment of intangible assets, which lack physical presence, the accurate identification and delineation of the valuation object are still of paramount importance. It ensures that all relevant components associated with the asset are included in the valuation process, preventing the omission of critical elements that may affect the overall assessment. The valuation subject often consists of a comprehensive collection of tangible assets. It raises essential questions regarding what is included in the valuation and what is excluded. Furthermore, it is vital to distinctly categorize the particular assets within the overall object of valuation.

Issues often arise concerning classification, as tangible assets can range from land and buildings to infrastructure, industrial machinery and computer equipment. Additionally, distinguishing between tangible and intangible assets, as well as the overall business value, becomes necessary during this process.

For example, practical questions may emerge about whether a crane system or the foundation of a large machine should be classified as part of the building or attributed to the machinery category. Similarly, considerations must be made regarding how improvements made by tenants in rented properties are treated in the valuation context, as well as how outdoor facilities situated on leased land should be classified.

To facilitate clarity and consistency in these classifications, definitions are established within the context of Fixed Asset valuation. By adhering to these definitions, valuers can navigate the complexities associated with identifying and assessing Fixed Assets, ensuring that the valuation process is thorough, accurate and reflective of the true economic value of the assets involved.

2.2.1 Fixed Asset Definition

Fixed Assets encompass a range of tangible assets often categorized under various general terms. While broader classifications such as "Fixed Assets," "Tangible Assets," "Capital Assets" or "Material Assets" encapsulate multiple subcategories, more specific terms like "Property, Plant and Equipment," "Machinery & Equipment" or designate particular groups within this category (see Figure 2.3).

Fixed Assets are often regarded as capital assets, reflecting their long-term nature, particularly in capital-intensive industries such as oil, automotive and steel manufacturing, where significant quantities of Fixed Assets are maintained.

According to the IVSC, Tangible Assets are "Assets with a physical manifestation. Example include land and buildings, plant and machinery, fixtures and fittings, tools and equipment and assets in the course of construction and development."[21]

Furthermore, a tangible asset is characterized as having a finite monetary value and a physical form. Such assets can generally be exchanged for monetary value, although liquidity may vary. This contrasts with intangible assets, which possess theoretical rather than transactional value.

According to RICS, "Plant and Equipment" includes all tangible assets, excluding Real Estate, that are utilized for production, rental or administrative purposes and are expected to remain in use over a designated period.[22]

The IFRS classify property, plant and equipment as tangible items that are used in the production or supply of goods or services and are anticipated to be maintained for more than one accounting period.[23]

While numerous definitions and interpretations exist, in them "(tangible) Fixed Assets" will be equated with "property, plant and equipment." Moreover, for this book, the term

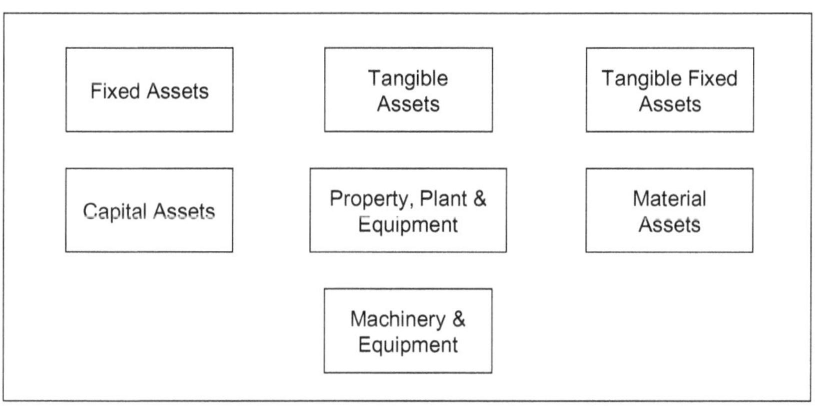

Figure 2.3 Fixed Asset-Related Terms.
Source: Authors.

Fixed Assets (property, plant and equipment) is divided into Real Estate (immovable) and Machinery & Equipment (movable). This broad term includes various asset classes, such as Real Estate (land and buildings) and M&E, along with fixtures, furniture technology and loose tools essential for operational effectiveness.

Heavy manufacturing i.e. the secondary sector, especially oil and gas, but also steel and automobiles, are classic Fixed Assets. It can be assumed that these companies have large amounts of these assets on their balance sheets.

The primary economic sector can also be very asset-intensive e.g. in mining. The tertiary sector i.e. the services sector, is traditionally not very asset-heavy. Here, Real Estate and office equipment are the main items in the Fixed Assets registers and balance sheets.

In order to be able to carry out a valuation of Fixed Assets, the valuers must always be familiar with the sector in question. In contrast to property valuation, where it is relatively unimportant what is produced in the property or which company occupies an office building, the valuation of Fixed Assets requires a very intensive study of the industry and the individual production processes of the company in order to determine the value of the Fixed Assets.

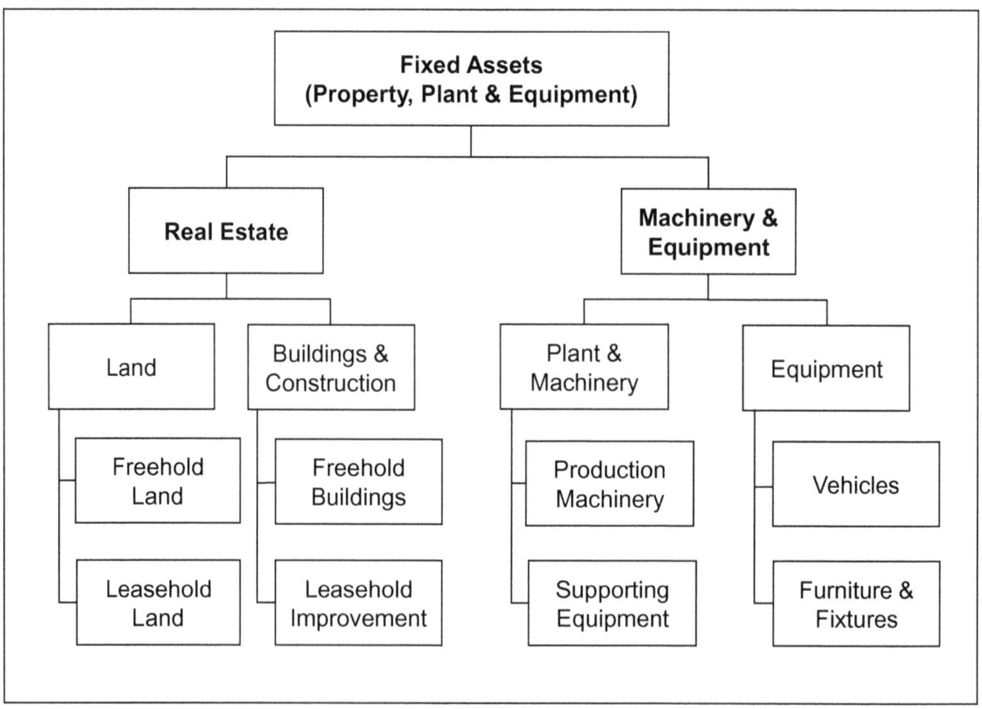

Figure 2.4 Fixed Asset Groups.

Source: Authors.

Note: Also see ASA [2011], pp. 24–30.

2.2.2 Real Estate

A single statutory definition of the term "Real Estate" does not exist.[24] However, a well-established broader definition is given by Schulte who defines Real Estate as: "assets which are economic goods comprising undeveloped or developed parcels, including the associated buildings and outdoor facilities. These assets are utilized by individuals within physical-technical, legal, economic and temporal constraints for purposes of production, trade, services and consumption."[25]

Another more economic definition is given by Grasskamp, who describes Real Estate as "an artificially bounded space possessing a fourth dimension, time – which is inseparably connected to the Earth's surface through the land. This space-time relationship is transformed within the Real Estate market into a money-time relationship, as the described space can provide specific utility within a given timeframe, meeting certain demand conditions."[26]

The Hong Kong Institute of Surveyors defines Real Estate as: "Land and all things that are a natural part of the land (e.g. trees, minerals) and things that have been attached to the land (e.g. buildings and site improvements) and all permanent building attachments (e.g. mechanical and electrical plant providing services to a building), that are both below and above the ground."[27]

In literature and Real Estate economics, the definition of Real Estate is also often approached through its different dimensions from where Real Estate can be described. As outlined in Figure 2.5 these are legal, physical and economical dimensions, which are affecting Real Estate.

Physical Dimension

This involves the examination of the building's material properties, a three-dimensional structure composed of walls, floors, ceilings and roofs, artificially delineating a segment

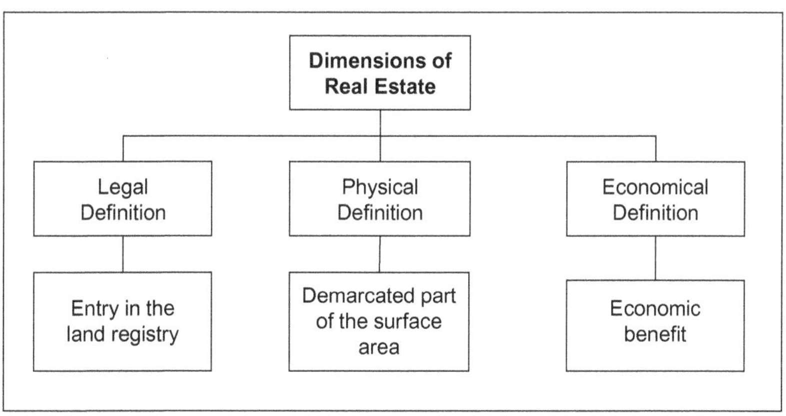

Figure 2.5 Dimensions of Real Estate.
Source: Authors adapted from Schulte et al. [2016], p. 5.

of the Earth's surface and the adjacent airspace. Critically, this dimension often excludes land considerations, focusing instead on substance-oriented definitions that prioritize utility generation over the physical object itself, especially in Real Estate investments.[28]

Legal Dimension

Although there is no formal legal definition for "Real Estate," it is viewed as a spatially defined portion of the Earth's surface (land and ground) along with permanently attached components and rights, forming an economic unit.[29] Theoretically, the conical volume from the Earth's center to outer space is considered part of the property. A constitutive element in most territories is the registration in a specialized land registry under a unique number.[30]

Economic Dimension

From a production theory perspective, Real Estate is considered a production factor, either as a means of production or a site for service delivery. Investment-wise, it serves as a capital asset, generating economic value through its use. The owner's right to the defined space allows for the transfer of Real Estate, providing utility to third parties over time and yielding income by converting space-time units into money-time units.[31] Thus, the economic nature is not derived from physical composition but from usage possibilities and market recognition. The value of Real Estate is determined by the market's appreciation of its utility.[32]

Characteristics of Real Estate and Real Estate markets

Real Estate is distinguished by several unique characteristics which also make its valuation much more complex than for e.g. commodities or shares which are frequently traded on international stock markets with prices available at almost any point of time. The characteristics of Real Estate are including but not limited to:[33]

- **Immobility:** Real Estate is fixed in location and cannot be moved, having the major impact on valuation and market behavior.
- **Heterogeneity:** Each Real Estate asset is unique in location, design and utility, contributing to market differentiation.
- **Limited Substitutability:** Real Estate cannot be easily replaced or exchanged, leading to unique investment dynamics.
- **Development Duration:** Developing Real Estate involves significant planning and construction phases.
- **Lifecycle Length:** Real Estate typically has a long lifecycle, enduring over extended periods and requiring periodic maintenance.
- **Investment Magnitude:** Real Estate investments require substantial capital, often necessitating significant financial commitments.
- **Transaction Costs:** Real Estate transactions involve high costs due to legal, regulatory and market factors.

Due to these special features of the Real Estate asset class, the Real Estate market also has certain special characteristics, which can be described as the following[34] :

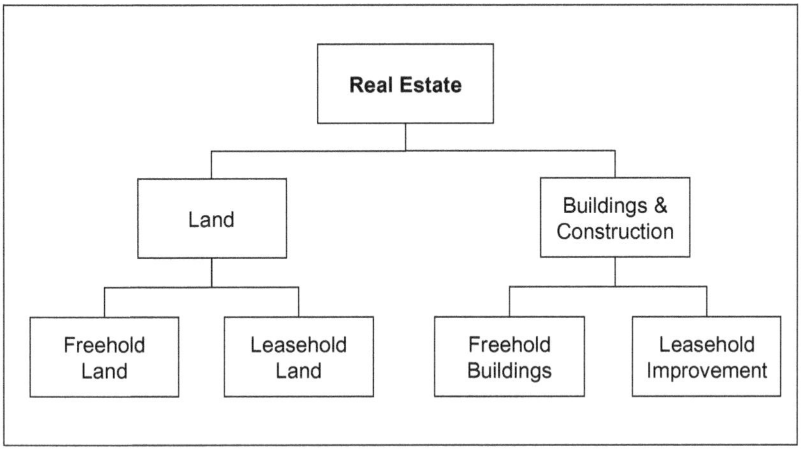

Figure 2.6 Real Estate Categories.
Source: Authors.
Note: Also see Floyd and Allen [2002], p. 17.

- **Spatial Submarkets:** Real Estate markets are divided into distinct geographic submarkets, each with unique characteristics.
- **Segmental Submarkets:** These markets are also divided by property type, with specific dynamics.
- **Insufficient Market Transparency:** Information asymmetry can be prevalent, complicating market operations and valuation accuracy.
- **Dependency on Other Markets:** Real Estate markets are interconnected with broader economic, financial and social systems, impacting their dynamics.
- **Low Elasticity to Market Changes:** Real Estate's fixed supply and demand constraints can lead to limited responsiveness to short-term market fluctuations.
- **Cyclicality:** Real Estate markets are subject to cyclical trends, influenced by macroeconomic factors and investor sentiment.

As shown in Figure 2.6, Real Estate can basically be differentiated into two groups: Land and Buildings & Construction, with each further including leasehold and freehold.[35]

In summary, Real Estate is a utility-generating economic asset, with its value determined by its market-recognized use and potential uses. Understanding these dimensions is critical for accurately assessing Real Estate's role and value within fixed asset valuation.

2.2.3 Land

Land is the fundamental asset within the realm of Real Estate, serving as the foundational element upon which all property is constructed. It encompasses various legal forms and development statuses, each with distinct implications for ownership, utilization and valuation. Due to the fact that land is tied to a specific location and the resulting characteristics of Real Estate, as described above, the location of a property is by far the most important value-relevant factor of a property. Simply put, you can change everything about a property, but not its location.[36]

Moreover, the legal status of land has an impact on the valuation as well. The legal classification of landownership typically falls into two primary categories: freehold and leasehold.[37]

- **Freehold**: This form of ownership confers indefinite rights to the land and any structures upon it, representing the most comprehensive form of property rights. Freehold ownership allows the proprietor to utilize, transfer or develop the land without temporal restrictions, subject to legal regulations.
- **Leasehold**: Leasehold ownership entails holding the land for a predetermined period, as specified in a lease agreement. During the lease term, the leaseholder possesses the right to utilize and develop the land, although ultimate ownership remains with the freeholder. Leasehold arrangements are prevalent in both commercial and residential Real Estate, where long-term leases are negotiated.

Since freehold land has an unlimited useful life, unlike other assets, freehold land is not depreciated in the accounting books. For leasehold land, besides the location, the remaining leasehold period plays a major role in the analyses of its value.

For leasehold land the three major characteristics to consider are the lease term, the agreed lease payments and the reversionary clause outlining the compensation the owner of the land has to pay for buildings and construction built by the lessor. Those clauses are typically laid out in the leasehold contract analyzed by the value and/or respective local regulations.

Besides its legal status, the second characteristic of land is its development status. Land can be categorized based on its development status, influencing potential uses and Market Value. The primary statuses include:

- **Raw Land**: Referring to parcels that remain untouched by development or construction, raw land encompasses natural landscapes such as forests, fields or wetlands. It is often sought for agricultural, conservation or future development purposes.
- **Improved Land**: Improved land has been enhanced with infrastructure, buildings or facilities, rendering it suitable for specified uses such as residential, commercial or industrial activities. The degree of improvement may vary widely, from basic infrastructure to fully developed sites.

The legal development status of undeveloped land can vary significantly, particularly concerning its development potential, specifically regarding the type of permitted construction and the extent of allowable usage. This legal status has a substantial impact on the value of the land and is considered a critical factor influencing value, alongside location. Generally, the more valuable the potential economic use of the land, the higher its value will be.

2.2.4 Buildings and Construction

As with land, a legal distinction is also made between freehold ownership and leasehold in the case of Buildings and Constructions. However, the properties to be valued will usually be freehold assets, since it is usually the owner of the property rather than the tenant who has an interest in the value of the buildings. In the case of leasehold assets, buildings and constructions regularly involve so-called tenants or leasehold improvements. These are structural changes or fixtures that the tenant has made to the building they have rented. Since the tenant normally owns such leasehold improvements, they are to be examined as part of a valuation and only taken into account if tenants' owned assets are in the valuation scope, but in valuation practice they usually only play a minor role.

When analyzing and valuing buildings and constructions, the distinction that is regularly more important and influences value in practice is the differentiation according to the respective type of use.

The characteristics of Real Estate, above all its immobility, as well as the use function of Real Estate, lead to a segmentation of the Real Estate market into numerous submarkets.[38] In addition to a geographical classification (immobility), a functional systematization (usage function) of the various properties, known as property types, has become established. As a general rule, as shown in Figure 2.7 below, a distinction is made between residential, commercial, industrial and special-purpose properties, each of which includes various subcategories.[39]

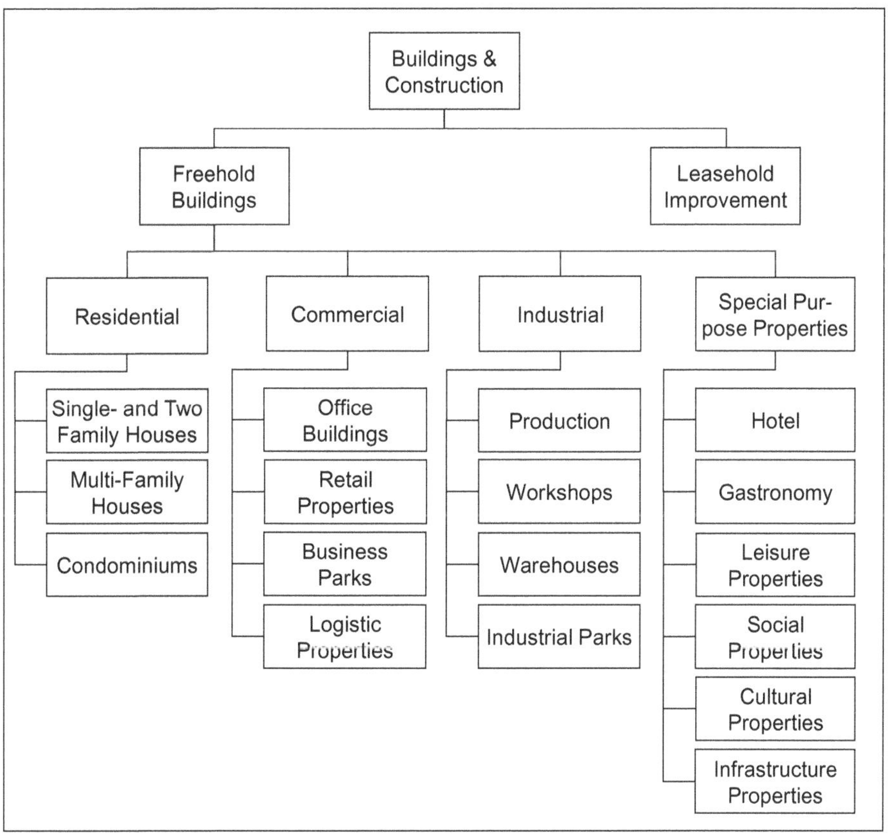

Figure 2.7 Types of Real Estate.

Source: Adapted from Schulte et al. [2016], p. 85.

Notes: Also see Fahrländer and Kloess [2024], p. 53.

- **Residential Properties** comprise several types, each serving different housing needs. Single and two-family houses are designed to accommodate individual families or up to two families, offering private residential spaces typically with their own land. Multi-family houses and residential complexes include buildings with multiple housing units, often occupied by several families or individuals. These can range from small apartment buildings to expansive residential complexes. Condominiums represent individual residential units within a larger building or complex, where each unit is owned separately while common areas are jointly owned by all unit owners.

  Residential properties generally play a very important role in property valuations and are a central component of many valuation projects. However, when property valuations are carried out in connection with the valuation of M&E, the situation is usually different, and residential properties play a subordinate role. In this case, commercial property, in particular industrial properties including production buildings, workshop, storage and warehouse facilities as well as office properties, play a significant role.
- **Commercial Properties** serve various business functions and are categorized into several types. Office properties are designed for business use and include spaces for administrative and professional services, ranging from small office buildings to large corporate headquarters. Retail properties are used for commercial retail purposes, comprising shopping centers, boutiques and other retail outlets where goods and services are sold to consumers. Business parks are specially designed areas that house multiple businesses, often providing shared facilities and infrastructure tailored for commercial enterprises. Logistics properties include warehouses and distribution centers, essential for storing and managing goods in transit to various destinations.
- **Industrial Properties** focus on manufacturing and production activities. Production buildings are facilities specifically designed for manufacturing processes, accommodating machinery and production lines. Workshops serve as smaller-scale industrial spaces used for crafting, repairs and small manufacturing processes. Warehouses are large buildings designed for the storage of goods, often integral to logistics and supply chain operations. Industrial parks are designated areas for industrial businesses, offering infrastructure and facilities suitable for manufacturing, storage and distribution activities.
- **Special Purpose Properties** encompass a diverse array of functions. Hotel properties provide accommodation services for travelers and tourists, ranging from small inns to large hotels. Gastronomy spaces are dedicated to food service and dining establishments, including restaurants, cafes and bars. Leisure properties are used for recreational activities, such as sports facilities, amusement parks and entertainment venues. Social properties include facilities that provide social services, such as schools, hospitals and community centers. Cultural properties are buildings used for cultural purposes, such as museums, theaters and galleries. Infrastructure properties include essential public infrastructure, such as transportation hubs, utilities and telecommunications facilities.

Each category serves distinct functions and demands different considerations regarding investment, development and management.[40] Specialized properties are often so-called management properties; their value is closely linked to the operating performance of the management and the capabilities of the management to successfully manage the property. This has to be considered in the valuation.

In addition to the asset group Real Estate, which is characterized by its immobility, the second asset group consists of the movable Fixed Assets, the Machines and Equipment.

2.2.5 Machinery and Equipment

In the literature, there is neither a uniform definition of the terms M&E, also often referred to plant and equipment, nor a clear and consistent classification of the associated economic goods. In practice, it is common to utilize a structured hierarchy that categorizes M&E according to their respective industries.

The Royal Institution of Chartered Surveyors (RICS) defines all Fixed Assets which are not real property as "Plant, M&E " and describes it as the following[41]:

- **Plant:** This category encompasses assets that are combined with others and includes items that are part of industrial infrastructure, utility services, building services installations, specialized buildings,[42] as well as M&E that form a dedicated assembly.
- **Machinery:** This classification refers to individual machines, a collection of machines or a fleet/system of machines and technology. This can also include mobile assets such as vehicles, rail equipment, shipping vessels and aircraft. Machinery is defined as an apparatus designed for a specific process in relation to a user's industrial or commercial activities.
- **Equipment:** An overarching term that includes additional assets such as miscellaneous machinery, tooling, fixtures, furniture and furnishings, trade fixtures and fittings, as well as various types of equipment and technology. These items support the operational activities of the enterprise or entity.

Since there are no uniform and clear classifications available, for the purpose of this book, the (movable) Fixed Assets apart from Real Estate are categorized as outlined in Figure 2.8. The American Society of Appraisers (ASA) further categorizes M&E assets into additional subcategories, which have been adopted in the illustration as well.

The IVSC defines Plant and Equipment as Tangible Assets, other than Real Estate, that: are "held by an entity for use in the production or supply of goods or service, for rental by others, or for administrative purposes, and are expected to be used over a period of time."[43]

Depending on the economic sector or industry, a company may hold a diverse array of Fixed Assets, which can also include such assets in the course of being erected or constructed accounting for Construction in Progress (CIP) also referred to as Assets under Construction (AUC).

Understanding these different classifications as outlined is crucial for stakeholders involved in the valuation and management of M&E. By recognizing the different groups and subgroups, valuers and business owners can more effectively navigate the complexities of asset valuation e.g. allocation of total useful life to similar assets ensuring accurate assessments that reflect the true economic value of their M&E holdings.[44] The following classes and subclasses refer to the categorization made by the ASA[45]:

Construction in Progress (CIP)

Often referred to as work in progress (WIP) or Assets under Construction (AUC), Construction in Progress (CIP) designates projects that are currently undergoing construction and are

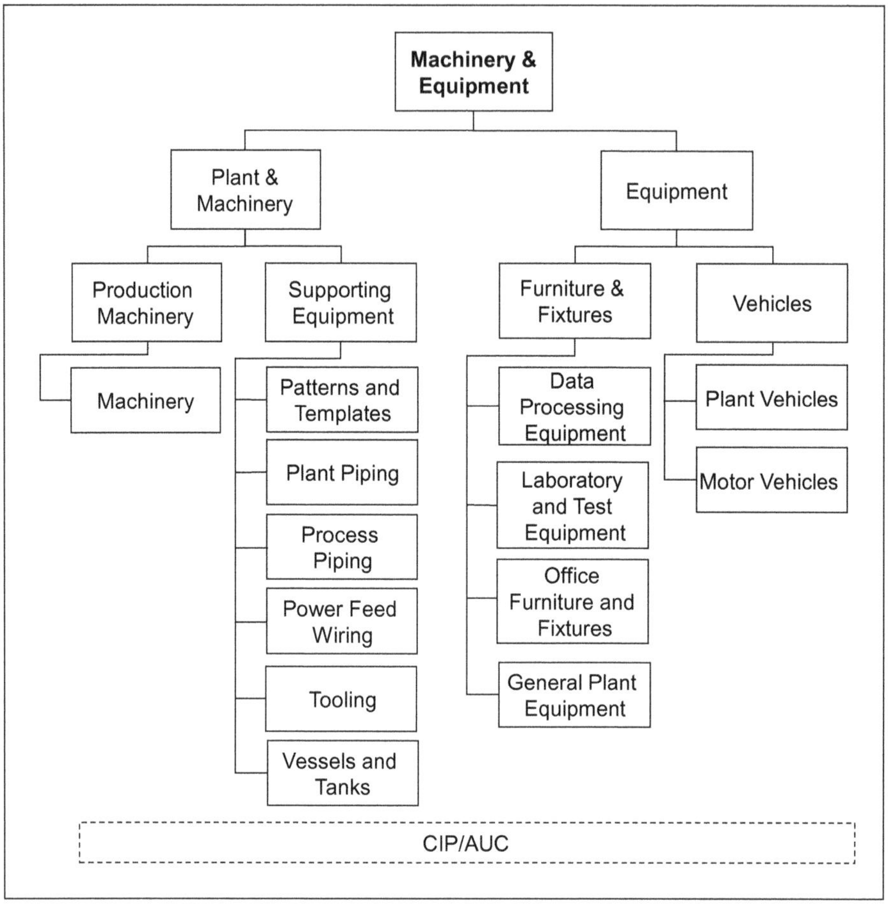

Figure 2.8 Machinery and Equipment Asset Classes.
Source: ASA [2011], p. 62.

reflected as capitalized assets within the owner's financial records. Such projects have not been completed as of the effective date of the appraisal.[46]

Data Processing Equipment

Data processing equipment encompasses a range of devices and tools utilized for the management and manipulation of digital information. This category includes computers, printers, plotters, terminals, modems and monitors, as well as word processing software. Additionally, it may also comprise other office machinery and devices, such as photocopiers and dictation equipment.

General Plant Equipment

This category encompasses lower-cost items that are vital for the operational efficiency of the plant. It includes elements such as factory furniture and fixtures, benches, racks, lockers, scales, hand trucks, time recorders, ladders, fire extinguishers and similar apparatuses.

Laboratory and Test Equipment

Refers to all equipment necessary for the effective functioning of laboratory or test facilities. Common items in this category include microscopes, clean tables, fume hoods, ventilation systems, spectrographs, ovens, stills, glassware and various other scientific apparatuses.

Machinery

Represents the most substantial assets in terms of personal property within a production-oriented business. This includes individual machinery units, manufacturing process systems and essential components such as wiring, piping and foundations integral to the production process.

Office Furniture and Fixtures

Encompasses items such as desks, tables, chairs, credenzas, filing systems, cabinets, safes and portable partitions – these furnishings are essential for office environments.

Patterns and Templates

This includes all patterns and templates employed in production processes, playing a crucial role in manufacturing consistency and quality.

Plant Piping

The term "plant piping" pertains to non-process industrial installations, covering components such as pipe fittings, valves, hangers, instruments, meters and other related fittings essential for operational integrity.

Plant Vehicle

Refers to specialized vehicles such as lift trucks, cranes and tractors that are not licensed for use on public roads yet are critical for intra-plant logistics and maneuverings.

Power Feed Wiring

Includes all electrical components necessary for the distribution of energy to electronically operated equipment within the facility, such as wiring, conduits, switchgear, distribution panels, circuit breakers, safety switches, bus ducts, bus bars, junction boxes and transformers (excluding those owned by the power company). This category may also incorporate substitution equipment owned by the client, along with concrete foundations and platforms required to stabilize the equipment, as well as fencing that provides safety enclosures.

Process Piping

The term "process piping" applies specifically to the piping systems utilized within manufacturing processes such as oil refineries, food production facilities and paper mills, where the transport of materials is critical to operational success.

Motor Vehicles

This category includes automobiles, trucks, tractors and trailers that possess the necessary licensing for road usage and are incorporated into the operational framework of the business.

Tools

Tools can be classified into two categories: permanent and perishable. Permanent tools are designed for extended use, often including portable electric tools, air tools, anvils, vises, gauges and chucks. Conversely, perishable tools are typically consumed during operation rather than worn down, comprising items such as chisels, reams, drills, collets and taps.

Vessels / Tanks

Encompasses tanks and vessels utilized in production settings, which are essential for the storage and processing of liquids and gases in various industrial applications.

The asset categories shown above are not exhaustive. There may be additional categories of assets that need to be classified and structured separately, particularly in specialized industries. Nevertheless, the assets shown are the typical groups found in the valuation of Fixed Assets.

2.3 Value Concepts and Value Definitions

As famously highlighted by Warren Buffet: "Price is what you pay, value is what you get,"[47] although the terms "value" and "price" are sometimes used interchangeably in daily language, they must be clearly distinguished from each other.[48] The price is the result of a concrete transaction and therefore always characterized by the individual and subjective notion of the parties to an asset. In contrast, the value of an asset is intersubjective and is therefore considered to be an objective price[49] based on an estimate reflecting the aggregate price expectations of a group of market participants.[50] It should be noted that there is not only one correct value of an item.[51] Under different valuation purposes, values can differ a great deal because of their definitional properties and consequent exposure measures.[52]

In the valuation process, there are various value concepts that arise from the different valuation reasons and the associated value definitions.

Depending on the valuation concept used, the value determined for one and the same valuation object can vary greatly. For example, the mortgage lending value may differ from the market value, and the market value may differ from the liquidation value. Therefore, knowledge and, above all, a clear definition of the determined value in the valuation report is essential. In addition, the condition of the valuation object can influence the derived value (e.g. Minimum Value, Scrap Value).

When determining the value, all actual, economic and legal circumstances that can influence the value of the asset must be taken into account.

Every valuation presupposes that there is clarity about the value to be determined. Different definitions of values can not only differ considerably in value from one another but can also have a considerable influence on the valuation procedure and the appraisal procedure to be used.

In the international valuation literature, a range of distinct concepts pertaining to fixed asset value of fundamental importance are predominantly described, which are elaborated upon below:

- Market Value
- Fair Value
- Existing Use Value
- Alternative Use Value
- Mortgage Lending Value
- Liquidation Value
- Minimum Value
- Scrap Value

Upon examining these value concepts, one can discern their close association with the specific purpose of valuation and the underlying methodologies employed in the valuation process.

Additionally, there are other terms related to value that must be distinctly differentiated in the context of costs, including:

- Original Cost/Historical Cost (Gross Book Value (GBV))
- Net Book Value (NBV)
- Replacement Cost New (RCN)
- Reproduction Cost New
- Depreciated Replacement Cost (DRC)

The most important value terms are defined and explained below. In addition to value measures and valuation premises, there are a large number of value terms, which are well reflected. The individual value definitions and cost terms are preceded by a short paragraph on the conceptual distinction between value and price.

2.3.1 Market Value

The Market Value has emerged as a central concept in the international Fixed Assets valuation practice, firmly established through the definition provided by the IVSC. This definition has been adopted verbatim by both TEGOVA (in the Blue Book)[53] and RICS (in the Red Book)[54] within their respective regulatory frameworks. According to International Valuation Standards (IVS) Market Value is defined as follows:

> Market Value is the estimated amount for which an asset or liability should exchange on the valuation date between a willing buyer and a willing seller in an arm's length

> transaction, after proper marketing and where the parties had each acted knowledgeably, prudently and without compulsion.[55]

In this context, the term "estimated amount" refers to the most probable price that can be reasonably expected under normal market conditions. For the seller, this represents "the best price that can realistically be achieved," while for the buyer, it signifies "the most advantageous price."[56] Furthermore, the definition highlights that the parties involved consist of a "willing seller" and a "willing buyer," both of whom are neither overzealous nor coerced and are acting in accordance with prevailing market conditions.[57]

The notion of an "appropriate marketing period" is critical to this definition, as the timeframe required to successfully market and sell a property can vary based on the unique characteristics of the asset and the current market situation. Thus, a universally applicable timeframe cannot be established.

Another vital component of the definition is the principle of the valuation date, which stipulates that the determined value must reflect the current market conditions as of the valuation date, rather than at any earlier or later time. Moreover, the definition inherently considers ordinary business transactions, ensuring that the parties do not operate under a special relationship that could lead to a price level not representative of the market.

For most valuation purposes, the Market Value is the subject value that must be derived. Closely related to the Market Value is the Fair Value. These are sometimes even used synonymously.

2.3.2 Fair Value

Fair Value, a term rooted in accounting principles, pertains to the valuation of assets for the purpose of financial reporting valuation.[58] Specifically, it is associated with the estimation of asset values within the context of annual financial reporting. According to the definition adopted by the International Accounting Standards Board (IASB) in IFRS 13 and also referred to by the RICS,[59] Fair Value is defined as:

> the price that would be received to sell an asset or paid to transfer a liability in an orderly transaction between market participants at the measurement date.[60]

This definition emphasizes the importance of market conditions and voluntary participation in the transaction and also makes reference to the measurement date.

In the context of fixed asset valuation, the term "Fair Value" typically corresponds to Market Value.[61] However, it is essential to note that Fair Value is sometimes described as a more comprehensive concept and may, in some specific cases, diverge from Market Value. This divergence can occur when unique circumstances surrounding the asset or liability impact its valuation.[62]

The IVS further clarify the definition of Fair Value by referring to IAS 16: "The Fair Value of items of plant and equipment is usually their Market Value determined by appraisal. When there is no evidence of Market Value because of the specialized nature of the plant and equipment and because these items are rarely sold, except as part of a continuing business, they are valued at their depreciated replacement cost."[63]

This highlights that in situations where market data is insufficient, valuers may resort to using the depreciated replacement cost as a practical measure for determining Fair Value. In practice, Fair Value and Market Value lead to the same result in most cases.

2.3.3 Existing Use Value

The Existing Use Value corresponds to the Market Value assuming that the current use of the asset is maintained.[64] With regard to Real Estate, it is assumed that it can be sold on the market in its current use and that it cannot be used differently than it is at present for the foreseeable future.[65] It should be emphasized that potential increases in value that could be achieved by a change in the existing use are deliberately not taken into account when determining the Existing Use Value.[66]

2.3.4 Alternative Use Value

The Alternative Use Value is defined by TEGOVA as "the value of the property under a use other than the present one."[67] The Alternative Use Value records the result of assuming a highest and best use of the valuation object. This can be either the current use or a planned future use that arises from a legally permissible and economically realizable change of use, including a development or refurbishment measure.[68] The reference to the highest and best use implicitly establishes the link to the Market Value, whereby the difference lies in the consideration of potential development measures.[69]

2.3.5 Mortgage Lending Value

The Mortgage Lending Value as defined by the Red Book is the value of a property: "as determined by an expert making a careful appraisal of the future marketability of the property, taking into account its long-term characteristics, normal and local market conditions, its current use and appropriate alternative uses."[70]

Despite the cut-off date principle, the value should have a long-term character, in contrast to the Market Value, and should not reflect short-term positive market volatilities.[71]

2.3.6 Liquidation Value

In Fixed Assets valuation, Liquidation Value refers to the estimated amount that would be realized if an asset or a group of assets were sold on an individual basis. According to the IVS, Liquidation Value is defined as:

> liquidation value is the amount that would be realized when an asset or group of assets are sold from a liquidation sale, with the seller being compelled to sell as of a specific date. Liquidation value can be determined under two different premises of value [...].[72]

This value should encompass the costs associated with preparing the assets for sale as well as the costs of disposal. Hence, additionally, TEGOVA defines Liquidation Value as "the estimated amount that creditors or equity holders could reasonably expect to receive in the event of liquidation of the company's property, plant, and equipment, when sold piecemeal or in whole."[73] Liquidation Value can be assessed under two distinct premises which may have a quite significant impact of the derived value.

- **Orderly Transaction:** In this scenario, liquidation occurs within a typical marketing period, allowing for a strategic approach to asset sales aimed at maximizing returns.
- **Forced Transaction:** This situation involves a rapid sale of assets due to urgent circumstances, resulting in a shortened marketing period and typically lower realizable values.

Therefore several regulatory bodies and definitions differentiate between the terms Orderly Liquidation Value and Forced Liquidation Value depending on the assumed marketing period.[74]

However, in practice, it is often challenging for the valuer to accurately estimate the impact of those specific marketing/disposal scenarios on the Liquidation Value. Typically, flat reductions are used in such assessments; however, these should be well-justified to ensure their validity.

2.3.7 *Minimum Value*

The minimum signifies the defined threshold for assets that, despite being fully depreciated in the financial accounts of the company but still continue to be actively utilized and contribute to the operational processes of a business. The primary purpose of establishing a Minimum Value is to recognize the ongoing economic utility of these fully depreciated assets. While they may hold a book value of zero on the balance sheet, their continued functionality, such as manufacturing equipment, machinery or vehicles, demonstrates their importance in supporting business operations and generating revenue.

In practice the Minimum Value is often applied at 5% or 10% of the assets related replacement or reproduction cost new. Depending on the industry and specific asset, this value can also be higher.

2.3.8 *Scrap Value*

Scrap Value, also known as residual or salvage value, represents the estimated monetary amount an asset can fetch at the end of its useful life after depreciation. It reflects the asset's worth based on recyclable materials or alternative uses and is vital for calculating depreciation, as it determines the depreciable amount by subtracting it from the initial cost, thus ensuring accurate financial statements. Influenced by market conditions, asset conditions and regulatory factors, Scrap Value provides strategic insights for asset replacement, investment planning and lifecycle management, helping organizations optimize capital expenditure and align asset strategies with financial goals.[75]

2.3.9 *Historical Cost (Gross Book Value)*

Historical Cost or acquisition or production costs refers to the original costs incurred at the time of acquisition or production of an asset. This includes both the purchase price of the M&E, as well as any incidental costs necessary to bring the asset into operation for the first time. In accounting terms, the historical cost is reflected by the GBV.

IFRS defines historical cost as: "The historical cost of an asset when it is acquired or created is the value of the costs incurred in acquiring or creating the asset, comprising the consideration paid to acquire or create the asset plus transaction costs."[76]

TEGOVA defines historical cost as "the cost of an asset when it was first placed into service by its first owner and was bought as new."[77] This definition underscores the importance of the initial expenses involved in acquiring the asset, which serve as a foundational measure of value in financial reporting and asset management.

The Historical costs play a fundamental role in the valuation of M&E when applying the Trending Method and are thus further discussed in detail in Chapter 6 of this book.

2.3.10 Net Book Value

Net Book Value (NBV) represents the value of an asset after deducting accumulated depreciation, amortization or impairment costs associated with that asset from its original cost. It's the amount at which an asset is reported on the balance sheet for accounting purposes. Hence the IVS defines Net Book Value in relation to a specific asset as "The capitalized cost less accumulated amortization or depreciation as it appears on the books of account of the business enterprise."[78]

The NBV is a key accounting measure used to assess the current carrying value of Fixed Assets on a company's balance sheet. NBV reflects the remaining value of an asset that has not been expensed or removed from the financial records, providing stakeholders with insights into the asset's historical investment and its current worth. While NBV is useful for financial reporting and internal management purposes, it is important to note that it may not always represent the asset's Market Value, as market conditions and other factors can significantly affect the actual realizable value of the asset if sold. This is the reason why e.g. in a Purchase Price Allocation exercise all identified NBV are replaced by its respective Fair Value.

2.3.11 Replacement Cost New

Replacement Cost New (RCN) is defined as the cost required to replace an asset with a new and flawless version that meets current functional requirements and specifications at the valuation date. This definition encompasses not only the base value of the new asset but also all necessary ancillary costs, such as transportation, assembly, installation and foundation work.[79]

The Replacement Cost New, is typically derived in the cost approach-based valuations for buildings and for M&E using the indirect cost approach when applying the trending method, the reproduction cost of an asset is important.

2.3.12 Reproduction Cost New

Reproduction Cost New, on the other hand, refers to the cost associated with creating an exact replica of an existing asset, using the same materials and construction methods utilized in the original. This valuation approach emphasizes the historical aspect of the asset and can be particularly relevant for unique or specialized assets that are not easily substitutable.[80] As mentioned, when valuing M&E assets based on trended (indexed) historical cost the derived result will be Reproduction Cost New.

2.3.13 Depreciated Replacement Cost

Whether it is a concept or a method has not been conclusively clarified in the literature.[81] In the Red Book of the RICS, DRCs are therefore also defined as "the costs for the restoration or replacement of assets, less deductions for obsolescence and all other relevant impairments and optimizations."[82]

2.4 Valuation Scenarios and Assumptions

A valuation exercise is always subject to a series of concepts and assumptions, which can have a significant influence on the scenarios to be assumed, the valuation approach and also the valuation results.

Since these conditions influence the entire valuation procedure, but also individual valuation methods and valuation parameters, they must be determined in advance of the valuation.

Not all assumptions can be made for every valuation occasion. It is therefore important that the fundamental assumptions underlying the valuation, and the resulting approach are agreed with both the client and other stakeholders involved in the process, such as auditors or tax authorities, and are then presented transparently in the valuation report.

In the following the main concepts and assumptions frequently applied in Fixed Assets valuation are explained.

2.4.1 Date of Valuation

The principle of the "valuation date" is a foundational concept in asset valuation, emphasizing the critical importance of determining an asset's value at a precise temporal point.[83]

Basically the date of valuation is important because the value of an asset changes over time. There are two reasons:

1 The condition of an asset changes over time.
2 The market environment in which the assets operate can change over time.[84]

Asset values are inherently variable, influenced by factors such as market trends, economic cycles, technological advancements and geopolitical developments. By establishing a clear valuation date, this principle mandates that valuations accurately capture market conditions and the intrinsic attributes of the asset as they exist on the specified date. Thus, the valuation date serves as a temporal benchmark, anchoring the valuation analysis within the specific economic, market and regulatory context prevailing at that time.

In addition to changes in market conditions that can affect an asset's value, the asset itself may undergo transformations over time that influence its valuation. These changes can be physical, economic or legal in nature.

For example, a machine may suffer physical damage or deterioration; a lease agreement, which significantly contributes to the cash flow of an office building, might be terminated by the tenant; or the legal use designation of a vacant parcel of land might be altered. Each of these events can substantially impact on the asset's value within a very short timeframe.

Consequently, the valuation date must be either in the past or coincide with the date of the physical site visit or inspection by the valuer. Given the unpredictable nature of future changes in the market or asset condition, the valuation date should never be set in the future (or if so, needs to be clearly stated and disclosed in the valuation report and documentation).

Adhering to the valuation date principle is often essential for compliance with legal and regulatory frameworks. Hence, accounting standards such as the IFRS and the US Generally Accepted Accounting Principles (US GAAP) require that valuations be conducted as of a specific date to ensure consistency and comparability in financial statements. This requirement upholds the integrity of financial reporting, allowing stakeholders to make decisions based on precise and timely information.

2.4.2 Going Concern

The going concern principle is a fundamental concept in accounting and financial reporting, applicable to both IFRS and national regulations. It is a crucial consideration outlined in IAS for the proper preparation of financial statements, assuming that an entity will continue its operations for the foreseeable future, typically assessed over a minimum period of twelve months from the reporting date.[85]

In most valuation practices, the going concern assumption is typically applied. This implies that all Fixed Assets are expected to maintain their current use as of the valuation date into the foreseeable future.[86] It is vital prior to conducting the valuation to ensure that the going concern assumption can be confirmed for all assets. This conclusion must also be clearly documented in the valuation report.

In cases where the going concern assumption is not applicable, the altered use of the assets, such as assuming the liquidation of the business and the subsequent sale of the assets, must be factored into the valuation. The value of assets can differ dramatically under a going concern versus a non-going concern scenario.

This is particularly pertinent when assessing specialized machinery or equipment whose value is mainly derived from the interaction with other assets e.g. in the context of a production plant. Without consideration of the current utilization, the value of these assets may not accurately reflect their worth in the context of a potential liquidation.

Hence, the going concern principle significantly influences Fixed Assets valuation, shaping the accounting treatment and measurement of assets.

2.4.3 Highest and Best Use

The principle of highest and best use is a foundational concept in Fixed Assets valuation. This principle is determined based on the optimal use of an asset as perceived by a typical market participant, even if the current use or intended use by the owning entity differs.[87]

The IVSC articulates the concept of highest and best use as follows: "The highest and best use is the most probable use of a property that is physically possible, legally permissible, and economically feasible and that leads to the highest value of the asset being appraised."[88]

TEGOVA similarly defines highest and best use as "the use of property, plant, and equipment (PME) that is physically possible, reasonably probable, legal or likely to become so, and that results in the highest value of the PME at the date of valuation."[89]

In practice, Fixed Assets are commonly utilized under their highest and best use.[90] However, there may be instances where the valuer identifies a potential for a more valuable use. In such cases, it sometimes becomes evident that the costs associated with transitioning the asset to this higher-valued use may exceed the potential increase in value that would result from the change. Consequently, the asset may be retained in its current use.

Applying the highest and best use principle is vital in Fixed Assets valuation, as it ensures that the valuer considers not only the current operational context but also the potential for maximizing the asset's value (e.g. what would a third party do with the asset). This principle allows for a more accurate and comprehensive evaluation of the asset's worth in the marketplace, reflecting the most advantageous utilization possible.

2.5 Purpose and Intended Use of a Valuation

The value of a fixed asset can be of interest for a myriad of reasons and purposes, necessitating an appraisal.[91] As stated by BVS [2024], "the foundations of valuation are significantly influenced by the occasion, mandate, and intended purpose of the report. Each valuation, and consequently each value derived from this criterion, is contextualized within this framework."[92]

When looking at the various valuation reasons, one can distinguish between two primary types of valuation occasions. The first category involves occasions where regulatory frameworks mandate a valuation, such as for tax compliance or financial reporting purposes. The second category encompasses instances where a valuation serves as an economic decision-making foundation, particularly in contexts such as investments, transactions, mergers and acquisitions or financing arrangements.[93]

Figure 2.9 allows an overview of the most relevant and frequently observed valuation purposes, which will be detailed in the following sections.

The occasion for the valuation significantly impacts not only the approach and methodology employed but also the nature and type of value to be determined, as well as the structure of the resulting valuation report,[94] Economic valuation occasions, which often function as internal decision-making tools, typically allow greater flexibility regarding the valuation scope and hence standards and methodologies that may be applied. In contrast, external valuation occasions, often driven by regulatory requirements, impose stricter guidelines on the valuer.

For instance, a preliminary value indication for internal decision-making related to economic or strategic business choices may be sufficient in the form of a brief presentation agreed upon with the client. Conversely, a valuation conducted for financial reporting purposes or to secure financing from a bank necessitates adherence to clear regulations and the delivery of a comprehensive report, which includes detailed inspections of the asset.

Thus, the occasion for valuation fundamentally influences the valuation process, the methodologies employed and the final deliverable valuation report. Consequently, it is imperative that these aspects are clearly defined and understood at the outset of the valuation engagement and clearly outlined in the engagement letter.

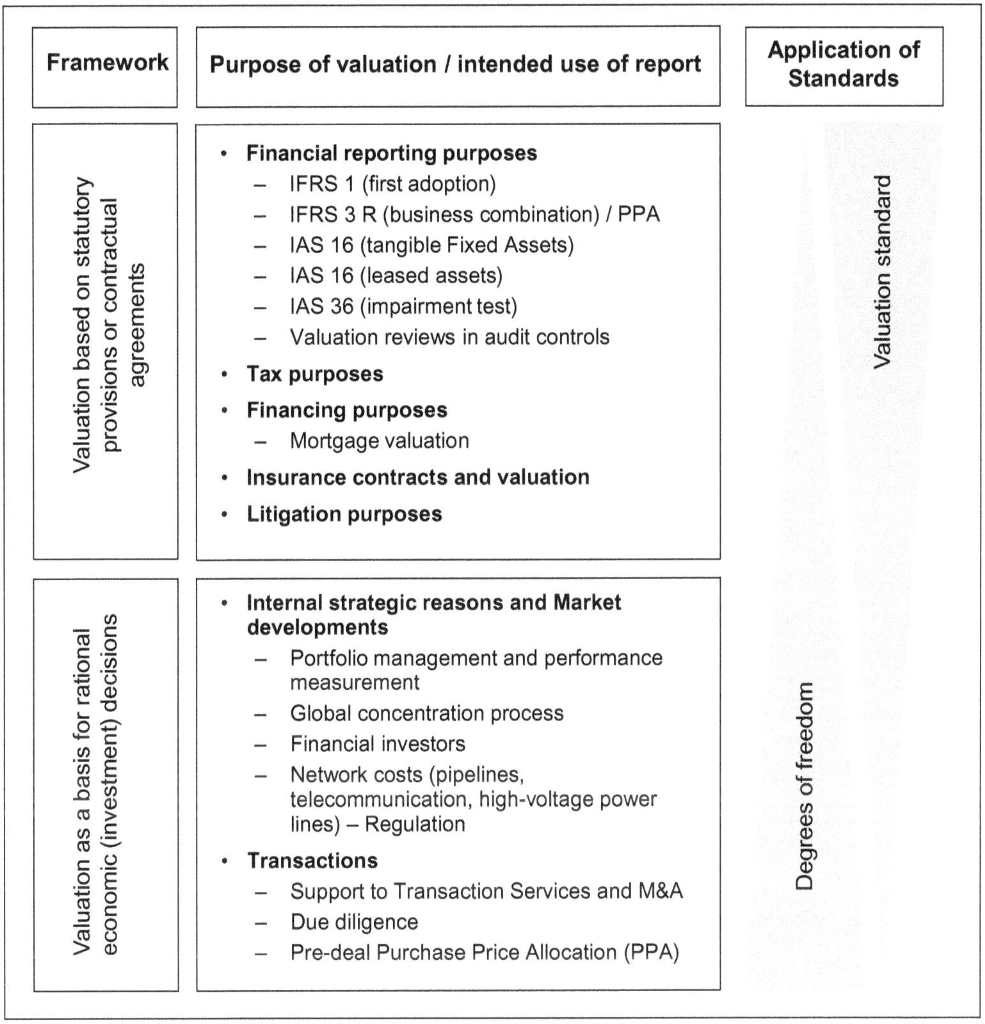

Figure 2.9 Valuation Purposes and Intended Use of Report.

Source: Authors.

Notes: Cf. Pfnür [2011], p. 50, Salden [2014], p. 28; Cf. Sommer and Kröll [2017], p. 2; Cf. Metzger [2018], p. 14.

The most common reasons for valuations and the special features that need to be taken into account are briefly described below.

2.5.1 Financial Reporting Valuation Purposes

The financial reporting valuation of Fixed Assets aims to provide accurate and reliable figures for the preparation of financial statements. This valuation must adhere to applicable accounting standards, such as the IFRS or local GAAP, depending on the jurisdiction. The primary focus is on Fair Value measurement, which reflects the price that would be received to sell an asset in an orderly transaction between market participants at the measurement date.[95] The valuation has substantial implications for the balance sheet and income statement.[96]

2.5.2 Tax Valuation

Tax valuation of Fixed Assets is essential for determining taxable value and ensuring compliance with tax regulations. This valuation must align with the tax code requirements of the relevant jurisdiction, which may significantly differ from financial reporting standards. It is important to differentiate between tax and book depreciation methods, as tax authorities often prescribe specific depreciation rates and methods.[97] Understanding available capital allowances and deductions that can affect taxable income is also crucial. Accurately valuing property for tax purposes may involve different valuation techniques compared to financial reporting. The primary impact of tax valuation is on a company's tax liability, affecting cash flow and potentially influencing investment and operational decisions.[98]

2.5.3 Insurance Valuation

Insurance valuation of Fixed Assets focuses on determining the appropriate level of insurance coverage required to protect against potential losses. This valuation ensures that the asset is neither underinsured nor overinsured. It involves deciding whether to insure assets based on their replacement cost or actual cash value. Additionally, the specific risks associated with the asset, such as location, usage and susceptibility to damage, must be evaluated. Understanding the terms and conditions of insurance policies, including any limits and exclusions that may affect coverage, is also critical. Insurance valuation affects the premiums paid and the potential recovery in the event of a loss, thus having financial and operational implications.[99]

2.5.4 Transaction Purposes

Valuation for transaction purposes involves assessing the value of Fixed Assets in the context of mergers, acquisitions or sales.[100] It is crucial for negotiating transaction terms and ensuring that the consideration exchanged reflects the true value of the assets involved.[101] Thorough due diligence is required to identify and assess all assets involved, ensuring that valuations are accurate and complete. Market comparisons and recent transaction data can provide valuable insights into valuation estimates. Additionally, considering the strategic value and potential synergies that the assets may bring to the acquiring entity is important. The valuation impacts the transaction price and terms, influencing the overall success and financial outcomes of the deal.

In each of these contexts, adherence to relevant valuation standards and methodologies is critical to ensure the accuracy, compliance and credibility of the valuation outcomes.

2.6 Valuation Framework and Guidelines

As outlined in the introduction already the valuation of M&E is not reflected in the literature as much as the area of Real Estate valuation. Hence while in many regions there are no detailed guidelines or legislation for the valuation of Fixed Assets or property, plant and equipment, there are several internationally applied guidelines and frameworks that provide guidelines for the valuation of Fixed Assets.

2.6.1 International Valuation Standards

At the international level, there are often no legal regulations regarding the valuation of Fixed Assets or at least real estate, which are sometimes established at the national level. Instead, professional associations and organizations are responsible for setting globally applied valuation standards and definitions, as well as ensuring their practical implementation.[102]

Within the domain of asset valuation, professional organizations serve a vital function in the establishment of standards, the certification of practitioners and the promotion of exemplary practices. Among the preeminent entities in this field, the RICS, the ASA, TEGOVA and the IVSC distinguish themselves through their substantial contributions to the discipline.[103]

The principles of valuation methods are defined in the "Red Book" of the RICS, which are internationally recognized and widely used in practice, particularly in the field of Real Estate valuation.[104] With regard to the harmonization of valuation at the European level, the "Blue Book" (European Valuation Standards) of TEGOVA represents a first step towards uniform valuation standards and is closely based on the content of the Red Book of the RICS.[105] In addition, the IVSC has existed since 1981 as a global association with the aim of creating global standards. The White Book (International Valuation Standards) contains valuation standards and procedural instructions for determining the value of assets for financial reporting purposes and is heavily influenced by the IASB.[106]

2.6.2 International Valuation Standards Council (IVSC)

The IVSC is an independent organization dedicated to the development and maintenance of international valuation standards. Its mission is to augment the consistency, transparency and reliability of valuations on a global scale, contributing to the stability of financial markets and fostering trust among investors. The IVS, promulgated by the IVSC, are widely recognized and adopted by valuation professionals worldwide, providing a common language and framework for valuation across disparate asset classes.

2.6.3 Royal Institution of Chartered Surveyors (RICS)

RICS is recognized as a leading global professional institution, accrediting experts within the sectors of land, property, construction and infrastructure. Renowned for its rigorous standards and ethical commitments, RICS offers a comprehensive framework for professionals engaged in the valuation of Real Estate and other asset classes. Its qualifications and certifications ensure that members adhere to elevated standards of practice. Furthermore, RICS plays a crucial role in the formulation of valuation standards that align with international best practices, thereby enhancing the credibility and reliability of valuations performed by its members.

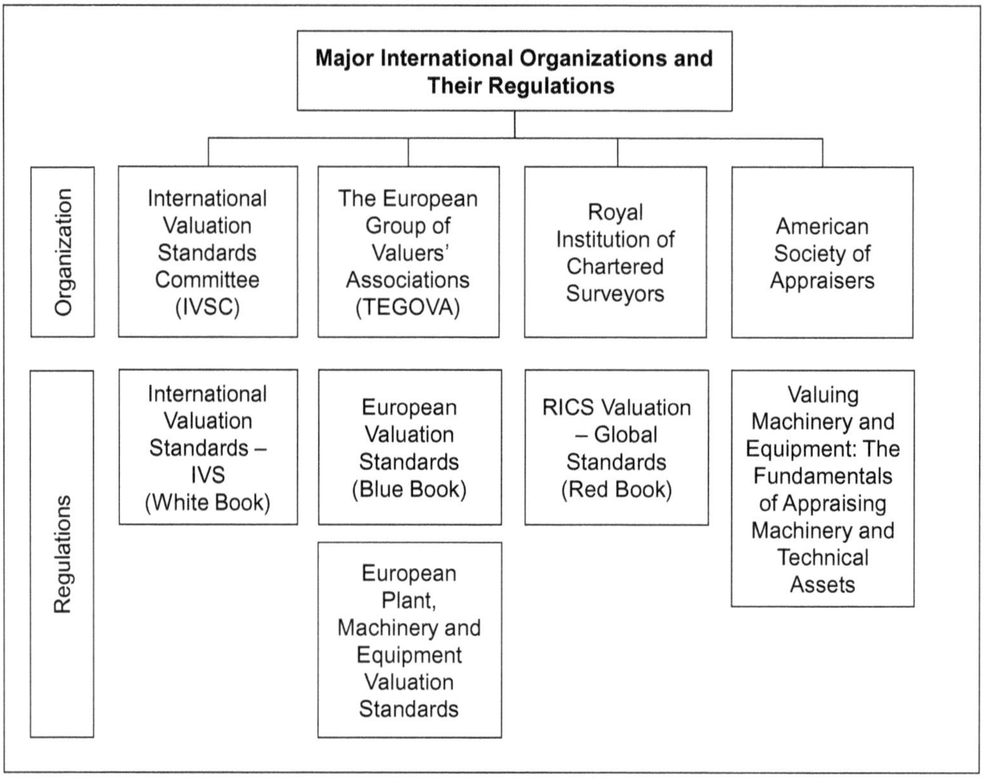

Figure 2.10 International Valuation Organizations for Fixed Asset Valuation.
Source: Authors.

2.6.4 American Society of Appraisers (ASA)

One of the most detailed guidebooks for the appraisal of M&E has been published by the ASA.

The ASA is a multidisciplinary organization representing appraisers across various asset categories, including Real Estate, M&E, personal property and business valuation. With a pronounced focus on education and professional development, the ASA provides certification programs that denote a high level of competency and expertise. The organization actively advocates for the appraisal profession and contributes to the evolution of valuation standards that accommodate the complexities and intricacies of diverse asset types.[107]

2.6.5 The European Group of Valuers' Associations

TEGOVA is an association encompassing professional valuation organizations throughout Europe, committed to the advancement of valuation standards and practices within the European Union and beyond. Its European Valuation Standards (EVS) offer guidance on valuation methodologies and ethical conduct, specifically tailored to the European market context. TEGOVA plays a pivotal role in harmonizing valuation practices across Europe, facilitating cross-border transactions and enhancing the transparency of the valuation process.[108]

These organizations, RICS, ASA, TEGOVA and IVSC, are instrumental in shaping the valuation profession by establishing standards, offering education and certification and advocating for ethical practices. Their endeavors ensure that valuations are conducted with integrity, precision and in accordance with globally acknowledged standards, thereby supporting informed decision-making in financial and investment contexts.

2.7 Financial Reporting Standards

The valuation of Fixed Assets for financial accounting purposes is intrinsically linked to the accounting frameworks that govern financial reporting. Among the most prominent frameworks are the IFRS, the Generally Accepted Accounting Principles in the United States (US GAAP) and various Local GAAPs that apply in specific jurisdictions. Each framework offers unique guidelines for the recognition, measurement and reporting of Fixed Assets, influencing their valuation in distinct ways.

2.7.1 International Financial Reporting Standards (IFRS)

IFRS, developed by the IASB, is a widely adopted framework that provides a global standard for financial reporting. In the context of fixed asset valuation, IFRS emphasizes Fair Value measurement and the revaluation model, allowing entities to reflect the current Market Value of assets on the balance sheet. The use of Fair Value can result in more volatile asset valuations, as it requires regular reassessment to capture changes in market conditions. IFRS also promotes transparency and comparability across international boundaries, facilitating cross-border investments and transactions. The framework's emphasis on Fair Value aligns with the needs of investors seeking up-to-date information on asset values.[109]

2.7.2 US Generally Accepted Accounting Principles (US GAAP)

US GAAP, established by the Financial Accounting Standards Board (FASB), is the predominant accounting standard in the United States. It outlines a more conservative approach to fixed asset valuation, primarily using the historical cost model. Under US GAAP, Fixed Assets are recorded at acquisition cost, with subsequent depreciation reflecting the asset's consumption over its useful life. The historical cost model provides stability and predictability in financial reporting, but may not reflect current Market Values, potentially resulting in understated asset valuations in times of significant market appreciation. US GAAP's focus on reliability and consistency appeals to stakeholders who prioritize these attributes over periodic market-based revaluations.[110]

2.7.3 Local GAAP

Local GAAP refers to the accounting principles and standards that are specific to individual countries or regions. These frameworks can vary significantly, reflecting local economic conditions, regulatory requirements and business practices. The treatment of

Fixed Assets under Local GAAP often falls somewhere between IFRS and US GAAP, with some jurisdictions permitting or requiring revaluation models similar to IFRS, while others adhere to historical cost conventions akin to US GAAP. Understanding Local GAAP is crucial for entities operating in specific markets, as it ensures compliance with domestic financial reporting standards and influences the perceived value of Fixed Assets in those contexts.[111]

Notes

1 TEGOVA [2022], p. 313.
2 Cf. Fahrländer and Kloess [2024a], p. 319.
3 Cf. Isaac and John [2013], p. 6.
4 Cf. Fahrländer and Kloess [2024a], pp. 319ff.
5 Cf. Adam [2018], p. 2.
6 Cf. Matysiak [2017], p. 1; Cf. Stand [2023], p. 9.
7 Cf. Stand [2023], p. 9.
8 RICS [2017], pp. 1ff.
9 Cf. Adam [2018], p. 2.
10 TEGOVA [2016], p. 1.
11 Cf. Adam [2018], p. 2.
12 Cf. Pfnür [2011], p. 47.
13 Damodaran [2005a].
14 Fahrländer and Kloess [2024a], p. 319.
15 Cf. BFH, Beschluss as of 22.5.2022, Az.: II R 61/99, in Kleiber [2007]; Cf. Fahrländer and Kloess [2024a], p. 319.
16 Also see Reinert [2021], p. 41.
17 Lost value can refer to e.g. transportation or installation cost.
18 Cf. Appraisal Institute [2004], p. 11.
19 Cf. HKIS [2017], p. 11.
20 Cf. O'Mara [1999], p. 144.
21 IVSC [2003], Glossary of Terms for IVS, p. 456.
22 RICS [2024], p. 124.
23 IAS 16, para. 6.
24 Cf. Renner and Sohni [2012], p. 32.
25 Cf. Schulte et al. [2016], p. 5; also see: Renner and Sohni [2012], p. 29.
26 Cf. Grasskamp [1972], p. 513.
27 Cf. HKIS [2017], p. 9.
28 Cf. Schulte et al. [2016], p. 5.
29 Hellerforth [2012], p. 166.
30 Cf. Schulte et al. [2016], p. 6.
31 Cf. Phyrr et al. [1989], p. 4.
32 Cf. Schulte et al. [2016], p. 8.
33 Cf. Schulte et al. [2016], p. 16; Cf. Brauer [2019], p. 6.
34 Cf. Schulte et al. [2016], p. 21; Cf. Brauer [2019], p. 9.
35 For further characteristics influencing the value of properties, also refer to White et al. [2003], pp. 21ff.
36 Cf. Fierz [2001], p. 28.
37 There are cases in which the topic of "mineral reserves" or "mineral rights" plays a role in the valuation of land. See Floyd and Allen [2002], p. 16. This occurs e.g. in the valuation of land with natural mineral resources such as mines. The topic of mineral reserves will not be discussed separately in the following sections; see Floyd and Allen [2002], p. 17.
38 Schulte et al. [2016], p. 16.

39 Schulte et al. [2016], p. 85.
40 Schulte et al. [2016], p. 105.
41 Cf. RICS Valuation Global Standards [2022], p. 11; Also see IVSC [2005], Guidance Note 3 Valuation of Plant and Equipment 3.1.
42 As to RICS certain specialized buildings can be categorized as Plant rather than real property or Real Estate.
43 IVSC [2005], Guidance Note 3 Valuation of Plant and Equipment 3.1.
44 ASA [2011], pp. 62.
45 ASA [2021], pp. 18ff.
46 As far as it has been checked that those costs were at market level, CIP are most of the time considered at book value. The book value is considered as a reasonable proxy for the Fair Value. It must be analyzed how long the CIP is already under construction. If the project got delayed or if the process took very long there might be an uplift or discount. Also, Real Estate Projects need to be considered at Market Value.
47 Warren Buffet.
48 Also see: Renner and Sohni [2012], p. 33; Schulte et al. [2016], p. 431.
49 There are also value terms that including certain subjective elements such as Residual Value or Mortgage Value.
50 Kleiber [2017], p. 454.
51 Cf. Fahrländer and Kloess [2024a], p. 320.
52 Metzger [2018], pp. 13ff.
53 Cf. "Standard 4. 10 - Bewertungsgrundlagen" in (TEGOVA [2003], p. 30).
54 Cf. RICS [2008], p. 10.
55 IVSC [2025], p. 24.
56 Practice Statement 3.2.1 Red Book p. 52; see also Parker [2016], pp. 17ff.
57 Practice Statement 3.2.4 u. 3.2.5 Red Book. p. 53.
58 Cf. Kleiber and Simon [2007], p. 495.
59 RICS [2024], p. 54.
60 IFRS 13, para 9.
61 Cf. Gondring [2004], p. 959; Cf. Meyer [2007], p. 138.
62 As an example, the RICS commentary (Practice Statement 3.2 Valuation principles, fair value) states that "fair value may reflect the synergistic value that would result from an exchange between two specific parties but would not be realizable in a sale in a wider market" (Red Book p. 57).
63 IAS 16, para. 32.
64 IVSC [2025], p. 28.
65 Cf. Petersen [2005], p. 30.
66 See also, TEGOVA [2003], p. 39.
67 TEGOVA [2025], p. 43.
68 Cf. Petersen et al. [2018], p. 60.
69 Cf. Gondring [2004], p. 960.
70 RICS [2008], p. 7.
71 Cf. Fahrländer and Kloess [2024a], p. 328.
72 IVSC [2025], p. 28.
73 TEGOVA [2022], p. 41.
74 Cf. IVSC [2025], pp. 28–29.
75 Cf. TEGOVA [2022], p. 41.
76 IFRS Conceptual Framework for Financial Reporting, Para 6.5, IFRS [2018]
77 TEGOVA [2022], p. 73.
78 IVS 2011, Glossary of Terms.
79 Cf. TEGOVA [2022], p. 318.
80 Cf. TEGOVA [2022], p. 319.

81 Cf. Hagen and Reif [2004], p. 31.
82 RICS [2008], p. 11.
83 Cf. TEGOVA [2022], p. 29; Cf. Kleiber [2018], p. 209.
84 Also see: Renner and Sohni [2012], p. 45.
85 IAS 1.25 and 126.
86 Cf. Fahrländer and Kloess [2024a], p. 327.
87 Cf. Fahrländer and Kloess [2024a], p. 326.
88 IVSC [2007] Cf. Also IVSC [2025], p. 28, p. 29; also see KPMG [2024], p. 77.
89 TEGOVA EVS, p 315.
90 TEGOVA EVS, p. 28.
91 Cf. Leopoldsberger et al. [2016], p. 427.
92 BVS [2024], p. 2.
93 Also see Metzger [2018], pp. 14ff.
94 Cf. Wyatt [2007], pp. 63ff.
95 Cf. KPMG [2024], p. 7.
96 Cf. IFRS [2018], p. 78; Also see IAS 40 [2000].
97 Cf. Australian Government [2022], p. 7.
98 Cf. Pfnür [2011], p. 52.
99 Cf. Salden [2014], p. 29, 33.
100 Cf. Manganelli [2015], p. 97.
101 Cf. Salden [2014], p. 29.
102 Cf. Thomas et al. [2000], p. 419.
103 Cf. Australian Bankers Association [2019], p. 5.
104 The RICS Valuation Standards, also known as the Red Book (so named because of its red cover), is the standard work of the British professional association, the Royal Institution of Chartered Surveyors. Even though the Red Book is only binding for the approximately 140,000 members of RICS worldwide when preparing written valuations, it also represents a widely recognized set of rules. It contains so-called "Practice Statements" (valuation standards) and more detailed "Guidance Notes" (explanatory notes). The Red Book does not contain any specific information on valuation methods and calculation methodology but is limited to the execution and reporting. (cf. RICS [2008]). In addition, the Red Book serves as a template and guide for the creation of local standards. Cf. HKIS [2017], Cf. NIESV [2015].
105 Cf. Petersen [2005], p. 51.
106 Cf. Thöne [2006], p. 6.
107 ASA [2021].
108 TEGOVA The European Group of Valuers' Associations.
109 Also see: The International Financial Reporting Standards, www.ifrs.org.
110 Also see: Financial Accounting Standards Board, www.fasb.org/.
111 Cf. e.g. EY [2022], IFRS versus German GAAP.

3 Preparation

While the basis of valuation establishes the fundamental framework within which the valuation project is performed, preparation marks the first phase of the valuation process itself (Figure 3.1). This stage can begin even before the official engagement or is sometimes integrated into the project proposal process to secure the valuation project. In this phase, the valuation project is planned and structured. It encompasses fundamental research and analysis aspects, wherein initial information, often publicly available, about the valuation object and the relevant market segment is gathered and analyzed. If the valuer is already engaged by the client, initial procedural steps, such as site visits and meetings, can be scheduled.

At the start of the valuation, it is crucial to gain a precise understanding of the valuation background, scope of engagement and the subject asset to be valued. These questions form the core content of this chapter.

As mentioned earlier, the delineation of each phase content is not always absolute, nor can activities always be strictly assigned to one phase. Moreover, the temporal sequence is not always linear and can, at times, proceed in parallel. It may also happen that elements from later phases are brought forward or vice versa. Depending on the individual project process, adjustments may be necessary. This applies e.g. to the planning and organization of site visits. While identification meetings may be planned during the preparation phase, final site visits often occur after the identification phase and the actual determination of production sites and facilities to be visited.

Nevertheless, the phases, along with the individual steps of the classical valuation process, provide a useful guideline for effectively structuring valuation process in practice.

At the outset of any valuation, the parameters of the valuation project must be clarified. The following key points, previously discussed as the basis of the value, are essential for the subsequent steps:

- What is being valued? – Nature of valuation subject and comprehensive understanding of the company's business as well as production process
- Why is the valuation being conducted? – Valuation framework and regulatory background

The answer to these two questions already sets the fundamental parameters for the entire further valuation process.

In the preparation phase, the commissioning of the work and further planning of the project are primarily conducted. The commissioning and the associated determination of the valuation scope are crucial components, which are detailed in the following sections.

DOI: 10.1201/9781003622413-3

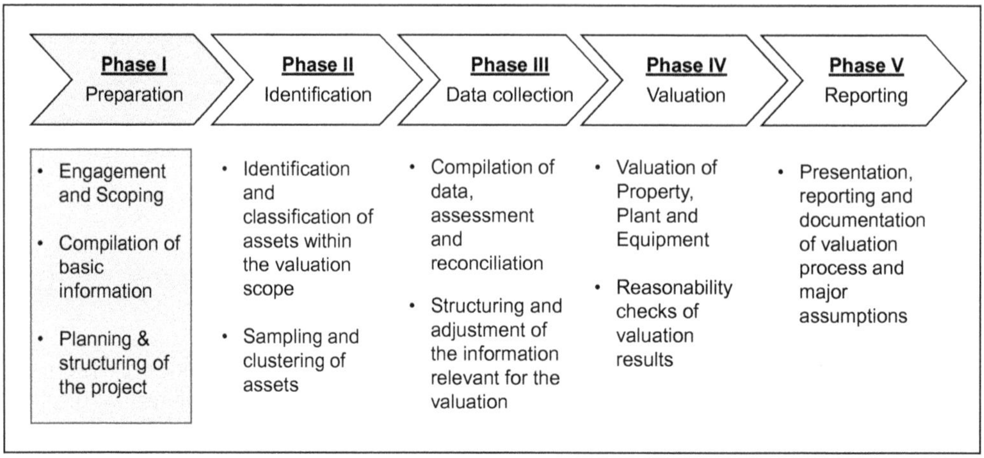

Figure 3.1 Five Phases Approach to Fixed Asset Valuation.
Source: Adapted from Plock et al. [2025], p. 20.

Initial information and data needed for these answers are obtained from the client Request for Proposal (RfP) or must be researched by the valuer.

3.1 Engagement and Scoping

The engagement and subsequently the engagement letter serves as the legal agreement between the valuer and the client, forming the essential foundation for the entire valuation exercise. This document outlines the scope of work, objectives and parameters of the valuation, ensuring alignment with the client's expectations and compliance with the valuation purpose as well as relevant legal and ethical standards.[1] By clearly defining the terms of engagement, the letter provides a framework for all subsequent valuation activities, facilitating transparency and minimizing potential conflicts of interest. It establishes the legal and procedural basis for the valuation, guiding the process and ensuring effective communication and collaboration between the valuer and the client. The following content is typically included in the engagement letter:[2]

- Name, address, contact person and contact details of the valuation firm and the client;
- Purpose of the valuation;
- Subject and scope of valuation;
- Valuation date;
- Scope of use of the valuation report;
- Due date and ways of submission of the valuation report;
- Total service charge or payment standard, payment due date and payment method;
- Other rights and obligations of the valuation firm and the client;
- Liabilities for breach of contract and dispute resolution clauses;
- Signing date of the contracting parties; and
- Venue where the engagement letter is signed or sealed.

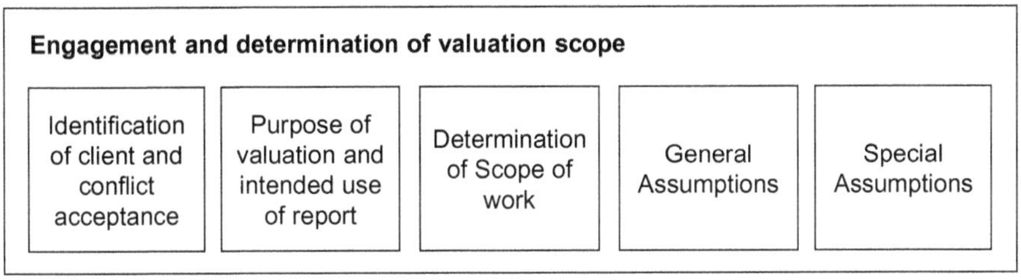

Figure 3.2 Engagement and Determination of Valuation Scope.
Source: Adapted from Appraisal Institute [2004], p. 51.

By clearly articulating these terms, the engagement letter fosters transparency and facilitates effective communication between the valuer and the client. This document not only guides the valuation process but also protects both parties by providing a reference point for resolving any disputes or misunderstandings that may arise during the course of the engagement.

Given that the engagement is the starting point of the valuation and all subsequent steps and activities are based on the terms outlined in the engagement, particular importance must be placed on the engagement and scope. Therefore, as depicted in Figure 3.2, several steps must be completed during the engagement process and definition of the valuation mandate before the engagement letter can be completed, signed and the actual valuation work can commence.[3]

3.1.1 Identification of Client and Conflict of Interest

The valuation process begins even before the client officially engages the valuer, specifically with the valuer's client acceptance process. The valuer must be able to clearly identify the client and later clearly name them in the valuation report. Before accepting the project, the valuer must conduct both internal and external checks (client acceptance check). External checks include anti-money laundering (AML) checks, which are typically mandatory in many countries e.g. also for Real Estate transactions. Internally, the valuer must ensure there are no conflicts of interest, allowing the mandate to be executed independently and without bias. Various organizations have independence regulations that their members must adhere to;[4] e.g. the Royal Institution of Chartered Surveyors (RICS), who have clear and strict ethical values for their members in their standards:[5]

> No member shall advise or represent a client where doing so would involve a conflict of interest or a significant risk of a conflict of interest, other than where all of those who are or may be affected have provided their prior informed consent. (The affected party can only give informed consent if the person explaining the position to them is entirely transparent, and also that the person explaining the position is sure that the party affected understands what they are doing – including the risks involved and any alternative options available – and is doing it willingly). Informed consent may be sought only where the member is satisfied that proceeding despite a conflict of interest is in the interests of all of those who are or may be affected.[6]

Essentially, RICS emphasizes that all potential conflicts of interest must be transparently disclosed and documented to avoid misunderstandings between client and valuer.

While it is impossible to compile an exhaustive list of all potential conflict of interest situations, the RICS further provides a helpful enumeration in its standards[7]:

- Acting for the buyer and the seller of a property or asset in the same transaction
- Acting for two or more parties competing for an opportunity
- Valuing for a lender where advice is also being provided to the borrower or the broker
- Valuing a property or asset previously valued for another client of the same valuer or firm
- Undertaking a valuation for third-party consumption where the valuer's firm has other fee-earning relationships with the client
- Valuing both parties' interests in a leasehold transaction.
- Members are also reminded that the interest of any third parties in the valuation, and the reliance they may place on it, will also be a relevant consideration.

There are situations where the same firm may support multiple bidders in a transaction. This can occur if, due to the size and complexity of the valuation scope, especially in an international context, only a few companies qualify as valuers, and some may already work as auditors for the company, which excludes them as valuers.

In such cases, transparency and disclosure of engagement relationships to the client are crucial, along with internal compliance.[8] This means that different teams within the company involved in the deal process must not share information and must adhere to strict confidentiality. This separation of teams within a company or organization is known as "Chinese Walls" or "Ethical Walls."[9]

3.1.2 Valuation Background and Intended Use of the Report

A crucial part of the engagement is understanding the valuation background and the intended use of the report. As discussed in Chapter 2, the reason for the valuation already provides insight into the value to be derived. For instance, if the valuation of Fixed Assets is intended to be performed for financial reporting purposes, a Fair Value will be determined. In contrast, if a company is being liquidated, the Liquidation Value will be relevant, whereas for securing loans, the lending value, and for insurance purposes, the insurance value, will be pertinent. Clearly identifying the value to be derived is essential within the preparation phase.

The valuation background and intended use of the report not only determines the value to be derived but also provides initial insights into the valuation scope and the necessary valuation report and relevant valuation date as well. For internal purposes, such as a potential transaction or internal decision-making, a short report of a few pages discussing the valuation methods, key approaches and results may be sufficient for the client, while for external purposes, a full scope valuation report is typically necessary. Typically, one can distinguish between valuations conducted for economic reasons and those for regulatory reasons. The latter has much stricter requirements regarding the minimum standards for valuation procedures and reports and are generally more elaborate than reports for internal purposes.

Based on the valuation background and intended use of the report, the valuer defines the scope of work to be performed in order to comply with the requirements for the valuation purpose.

3.1.3 Engagement and Determination of Scope of Work

As explained earlier, the engagement letter sets forth the responsibilities and expectations of each party, including the specific services to be rendered, the timeline for completion and the professional fee structure of the valuer.

One of the most relevant sections in the engagement letter is the scope of work.

The scope of work is defined and established in coordination with the client before the official engagement. This may necessitate involving other parties in the process, such as auditors, tax authorities, law firms or banks. For instance, it should be clearly determined in advance what level of valuation depth and documentation is required by the auditor or tax authority to ensure that the valuation report will be accepted and have the purpose of the valuation met.[10] It is crucial to clarify such requirements beforehand to avoid surprises later.

In practice, it sometimes happens that the client wishes to use the valuation report for purposes other than the originally agreed upon. This can lead to complications, particularly if the report is not fit for the new purpose intended by the client. For example, a report prepared for an internal purpose (e.g. internal decision-making) and later intended to be used for an external purpose (e.g. financing, financial reporting). The valuation depth or documentation in the valuation report prepared may be insufficient for such external valuation purposes, potentially placing the valuer in an additional risk position not accounted for in the original work scope or fee calculation. Therefore, it is essential to clearly describe both the scope of work and the intended use of the valuation in the engagement letter in order to manage the client's expectations and to protect the valuer.

As previously noted, defining the valuation object is a critical component of determining the scope of work. Although it may seem more straightforward for tangible assets than for business enterprises or intangible assets, this is a crucial step, especially in complex valuation project. This task can sometimes overlap with the identification phase. While individual assets are identified and clustered during the identification phase, the valuation object is defined and delineated during preparation. In early stages, it may be necessary to clarify the exact nature of the object to be valued, which often depends on the client's intentions and the background of the valuation.

As a result, as outlined in Figure 3.3, valuations projects are distinguished between Full Scope and Limited Scope Valuations.

3.1.4 Full Scope Valuation

The Full Scope Valuation can usually be considered as the standard in valuation engagements. Whenever a Fair Value, Market Value or Mortgage Value needs to be determined, a comprehensive full scope valuation, including inspection and assessment of all significant value-relevant assets, is necessary. This is typically the case for valuations driven by regulatory requirements, such as financial reporting, taxation or financing. The valuer requires thorough access to all information relevant to the valuation, and site visits and management meetings are conducted.

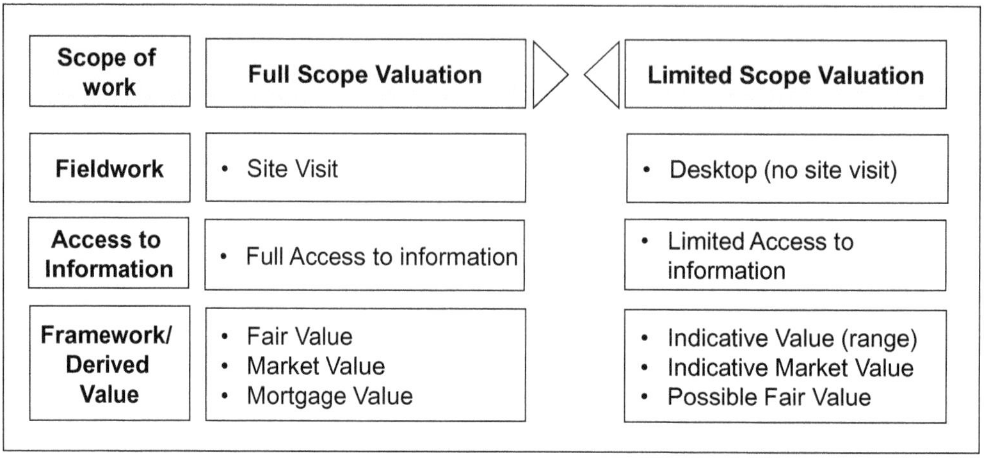

Figure 3.3 Scope of Work, Full Scope vs Limited Scope.
Source: Authors.

This does not mean that every single asset must be physically inspected and assessed for existence and functionality. In large asset portfolios, on-site inspections will focus on essential assets due to time and cost constraints. Since in many cases a few large machines and facilities often already represent a significant portion of a company's total tangible assets, a sampling approach may be employed. The same applies to Real Estate, although inspecting a building is generally quicker than detailing and assessing an entire factory's assets.

In certain cases, especially when not driven by regulatory requirements but by economic considerations, such as during an early stage of a transaction process an internal strategic decision-making process, a Full Scope Valuation may not always be necessary, and a more streamlined approach may be suitable to be applied. This is referred to as Limited Scope Valuation or Desktop/High-Level Valuation.

3.1.5 Limited Scope Valuation

The main distinction of a Limited Scope Valuation from a Full Scope Valuation is the restricted scope of work. Primarily, it foregoes on-site visits, with the valuation conducted from a desktop.

Since the valuer may not personally inspect the valuation object, they must rely on assumptions regarding the physical existence, condition and functionality of the assets. Typically, unless otherwise informed by the client, the valuation assumes full functionality and average condition of the assets.

Additionally, the valuer often has limited access to information, which can be a limiting factor in deriving a comprehensive Market Value. This limited approach is in most cases unsuitable for valuations required by regulatory frameworks.

This limited approach is rather used when a quick and preliminary value indication is needed to be produced and without extensive effort, often under time constraints. This is common e.g. at the beginning of a bidding process, where the seller uses an initial nonbinding offer to select potential buyers who will receive more detailed information as the process progresses. However, it is crucial to consider the valuation purpose and intended use of the report and ensure these are agreed upon upfront with all relevant parties involved.[11]

Additionally, a high-level desktop analysis involves numerous assumptions, which can significantly impact the valuation outcome. If assumptions deviate substantially from reality, the valuation result may be inaccurate. Therefore, it is essential to thoroughly explain all assumptions in the report and highlight any potential impacts on the valuation outcome, leading to a limited usability of the valuation report. In valuation it is typically distinguished between general assumptions and special assumption.

3.1.6 General Assumptions

In Fixed Assets valuation, general assumptions serve as foundational elements that underpin the valuation process. These assumptions are critical because they establish the context within which valuations are conducted and provide a common framework for analysis. While they can vary based on specific cases and contexts, some general assumptions are routinely applicable to ensure the reliability and validity of the valuation results. Moreover, certain assumptions laid out at the beginning of the valuation report serve as kind of a disclaimer for the valuer to clearly describe and point out under which conditions or limiting conditions the valuation has been performed.

Several common general assumptions are frequently seen during the valuation of Fixed Assets and considered in valuation reports:

- **Scope of Analysis**: Valuers often clarify that certain issues are outside their analysis scope. For example, they might state: "Environmental issues were out of our scope. Therefore, this analysis does not cover any environmental issues." This indicates that the valuation excludes potential environmental liabilities or compliance issues, focusing only on other factors. However; if valuers suspect environmental issues could significantly impact an asset's value, these must be explored and considered or at least outlined in the valuation report as a potential issue to be further assessed. This is particularly true for industrial properties with a long history of production and potential soil contamination.
- **Regulatory Compliance**: It is generally assumed for going concern that all necessary licenses, certificates of occupancy, consents or other forms of legal or administrative authority from relevant governmental or private entities have been obtained or renewed for any intended use analyzed in the report. This means that valuers assume all requisite regulatory approvals have been met, which supports the legitimacy of the asset's operation.
- **Improvements Compliance:** Valuers typically assume that all improvements made to the properties comply with the conditions set forth by statutory and other relevant legal, planning and engineering authorities. This assumption is crucial for ensuring that the enhancements to the asset do not violate codes or regulations, which could adversely affect its value.
- **Zoning and Legal Regulations**: Full compliance with applicable national, provincial and local zoning, their use, occupancy and environmental laws is generally assumed unless explicitly stated otherwise. This allows valuers to proceed with the analysis without scrutinizing the potentially complex legal frameworks governing the property.

- **Responsible Ownership and Asset Management**: The assumption of responsible ownership and proper asset management is fundamental, as it suggests that the asset has been maintained in a manner that preserves its value and utility. While this assumption is often present in valuation reports, the valuer should not be absolved of the responsibility to critically assess the maintenance condition of the Fixed Assets as part of their analysis. It is the valuer's professional duty to evaluate the condition of the valuation object and determine the implications for its value.
- **Operational Readiness:** In some instances, assumptions may be made regarding the asset's readiness for use. For example, it may be assumed that the facility is in a proper condition and equipped adequately to commence its intended operations without delay.

General assumptions are indispensable in the valuation process as they help simplify complex scenarios and facilitate a clearer understanding of the conditions impacting the asset being valued. Besides general assumptions, in certain valuation it may become necessary to also make certain special assumptions.

3.1.7 Special Assumptions

In the context of Fixed Assets valuation, special assumptions refer to specific scenarios or conditions that are established for the purpose of conducting an appraisal, particularly when the instructions provided differ from the actual circumstances present at the time of valuation. Hence, according to the European Group of Valuers' Associations (TEGOVA), a special assumption is defined as "an assumption made where instructions differ from the actual facts existing at the date of valuation."[12] The use of such assumptions is integral to the valuation process, often enabling valuers to provide relevant insights and estimated values reflective of unique situations or intentions behind the valuation.

Several circumstances may necessitate the formulation of special assumptions in Fixed Assets valuations, including but not limited to:

- **Market Conditions:** Valuers may encounter situations where the prevailing market conditions at the time of valuation do not align with the assumptions or expectations outlined in the appraisal instructions. For instance, if a property is being valued as if it were fully leased, but it is currently not generating income due to vacancies, the valuer may need to make special assumptions regarding the expected rental income or occupancy rates.
- **Intended Use of the Valuation:** Special assumptions are often required when the purpose of the valuation specified by the client diverges from the actual state of the asset. For example, if a fixed asset is being valued for potential sale under favorable conditions, but the current operational status suggests otherwise, assumptions may need to be made regarding potential enhancements or modifications to the asset that would facilitate its marketability.
- **Regulatory or Legal Considerations**: In certain cases, regulatory or legal frameworks may mandate specific conditions or restrictions that conflict with the existing facts. For instance, a valuation may need to account for zoning changes or pending regulatory approvals that could significantly impact the asset's future utility and value.
- **Assumed Economic Life**: The economic life of an asset may vary based on various factors such as technological advancements, market demand and operational

efficiencies. Valuers may need to make special assumptions about the expected life span of machinery or equipment, particularly when industry standards differ from the specifics of the asset being evaluated.

- **Valuation Date in the Future:** Under certain circumstances, a client may require a future valuation date for Fixed Assets. However, the valuation date should not be set in the future when determining a Fair Value due to potential changes in the valuation object or market conditions, affecting all definitions of value. This restriction is due to uncertainties and potential fluctuations impacting the asset and market environment. Economic shifts, technological advancements or regulatory changes can significantly alter the future value of assets. Thus, it is crucial to choose a valuation date reflecting current conditions and information for accuracy and reliability.

  If a value analysis with a date of value in the future is needed, it should be part of a forecast or scenario analysis, involving assumptions about future developments. These analyses differ from actual valuations as they rely on hypothetical assumptions and have different objectives.
- **Status of Development:** In some valuation scenarios, a special assumption may be applied, such as valuing an asset as if its development is complete. This assumption, known as "as if completed," allows the valuer to assess the potential Market Value of the property assuming all planned development, construction or improvements have been fully realized. This approach is particularly useful when a project is still under construction or in the planning stages, providing stakeholders with an estimate of the asset's value upon completion. However, it is crucial for the valuer to clearly communicate this assumption in the report, highlighting that the valuation is contingent upon the project's successful completion and does not reflect the current, incomplete status of the asset.

The incorporation of special assumptions in a valuation can introduce a caveat to the resulting valuation outcome. Consequently, by adopting such assumptions, the valuer is likely not determining a standard Market Value but rather a "market value under special assumption." It has to be noted that this distinction can lead to substantial deviations from a typical Market Value, potentially affecting the applicability and reliability of the valuation report, particularly in contexts involving external regulatory valuation requirements. It is imperative for valuers to clearly articulate these special assumptions and their implications to ensure that users of the valuation report fully understand the conditions under which the valuation is derived.[13]

3.2 Planning and Structuring

After the engagement process, or even in parallel with it, the valuer begins with the planning and structuring of the project. Project planning is a critical component in the valuation of Fixed Assets, providing a structured framework for the engagement while ensuring alignment of objectives, timelines and deliverables among stakeholders. Derived from the timeline agreed upon in the engagement letter, the project plan outlines procedures and considerations for planning and scheduling, with a focus on the identification meeting.

When examining the classical valuation process depicted in Figure 3.4, the preparation phase is concerned with gathering initial information to both organize and structure the project, as well as define the engagement and scope of the valuation.

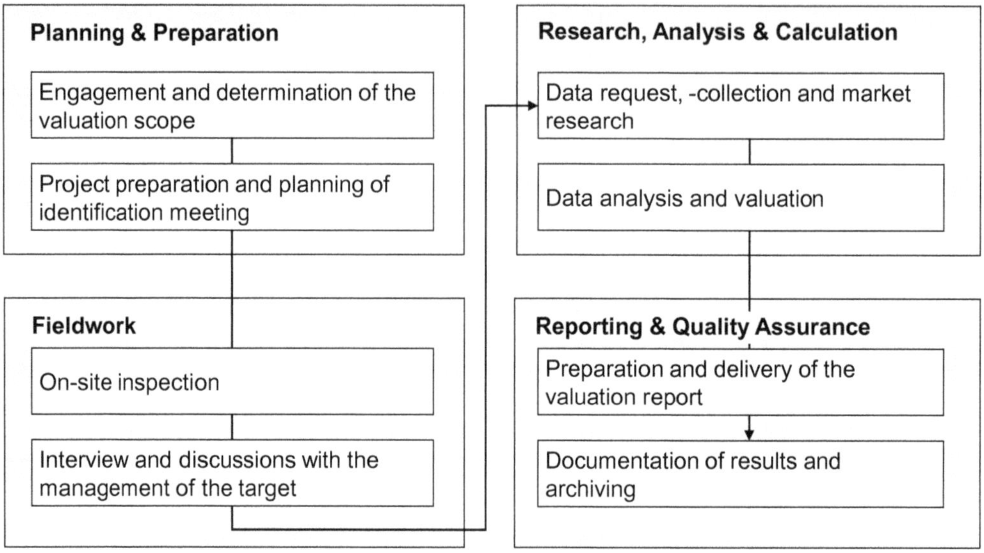

Figure 3.4 Valuation Process.
Source: Hackelberg and Hennig [2021], p. 326.

As illustrated in Figure 3.4, the individual steps are predetermined, ranging from planning and preparation to implementation and culminating in the revision of the report and documentation obligations.

3.2.1 Project Planning and Scheduling of Identification Meeting

The initial phase of project planning involves creating a preliminary timeline that acts as a roadmap for the valuation project, guiding it from inception to completion. Identifying key milestones and securing agreement with the client are essential for tracking progress and aligning with client expectations. Effective time management and scheduling are crucial to accommodate the scope and complexity of the project.

To visualize the project plan, a diagram typically illustrates the sequential distribution of project phases over the duration, aiding in understanding the timeline and dependencies. A typical project schedule is shown in Figure 3.5.

Interim project meetings are crucial for maintaining momentum and ensuring alignment. Projects generally start with a kickoff meeting to brief participants on the process and responsibilities, and finalize the project plan, including scheduling identification meetings, site visits and the final presentation. Consensus among stakeholders on the project plan is paramount. Should delays occur, such as in the provision of critical data, these are promptly communicated through regular update calls scheduled weekly or bi-weekly.

Effective project planning and scheduling are vital for executing a successful Fixed Assets valuation project. By establishing a clear timeline, coordinating key meetings and ensuring robust communication, the project team can navigate complexities and deliver

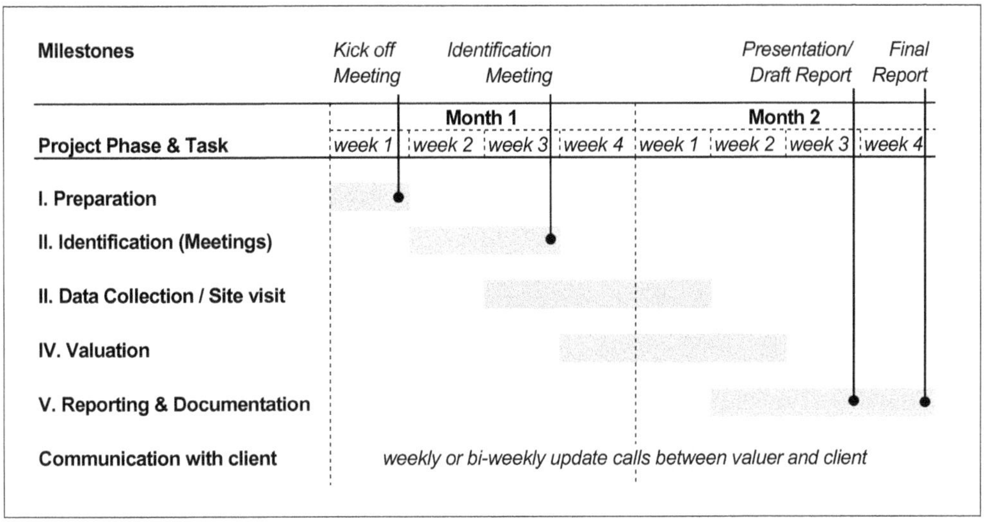

Figure 3.5 Project Plan and Milestones.
Source: Authors.

reliable outcomes. The identification meeting is central to maintaining focus on assets critical to the client's objectives.

3.2.2 Collection of Basic Data and Information

Although the actual data collection commences after the engagement and during the Data Collection Phase (see Chapter 5), some general basic data must be gathered at a very early stage.

This initial data collection includes information about the client and the valuation object, as well as the valuation background. These data are primarily needed to address the fundamental questions raised at the beginning of the chapter of "what is being valued" and "why is the valuation being conducted" as mentioned earlier.

Some of this information is often provided by the potential client in the form of a RfP, which is made available to all valuers involved in the bidding proposal process of the valuation project. In addition to this initial data provided, the valuer will conduct their own further research to gather additional essential information about the client and the valuation object.

Typically, the valuer will rely on publicly accessible information. This may often include the company's annual financial reports or online information about the client and their business activities.

This information collected prior to the engagement must be sufficient to conduct the client acceptance process, as described in the beginning of this chapter, without any doubt. It should also support informed project planning and the professional fee calculation. If sufficient information is not available at the time of the engagement, the valuer must clearly state this in the engagement letter and reserve the right to adjust both the timeline and fee structure during the project's course.

Notes

1 Cf. Renner and Sohni [2012], p. 36.
2 Cf. Chinese Valuation Standards [2013], p. 31.
3 The explanations are primarily concerned with the procedure for a valuation assignment. However, even in the case of a valuation review or valuation consulting project, individual areas, particularly with regard to client acceptance, should be carried out in the same way as for a valuation assignment, while others may need to be handled somewhat more freely.
4 Cf. HKIS [2017], p. 11.
5 RICS [2022], 3.2. p. 24.
6 RICS [2022], section 3.2. p. 24.
7 RICS [2024] section 3.9. p. 25.
8 Also see: Renner and Sohni [2012], pp. 36ff.
9 Cf. HKIS [2017], p. 25.
10 Cf. Australian Government [2022], p. 8.
11 Cf. Australian Bankers Association [2019], p. 3.
12 TEGOVA [2022], p. 319.
13 Cf. TEGOVA [2025], pp. 42 and 71; Cf. Australian Bankers Association [2019], pp. 2–3.

4 Identification

The identification phase, closely linked to the preparation and scoping phase, constitutes the second stage in the project timeline and serves as the foundation for subsequent valuation activities. Broadly speaking, this phase involves two key components:

1. the identification of assets to be appraised and
2. in cases where a larger number of assets are involved, the clustering and/or sampling of the asset portfolio.

Although the detailed identification phase formally begins after the engagement has started, preliminary identification often occurs already prior to engagement. This early assessment is critical for estimating the valuation scope and drafting the valuation engagement letter and, consequently, for calculating the valuer's professional fees.

During the identification phase, essential assets to be valued are precisely determined, and the valuation framework is defined. Often the financial accounts form the primary basis for this process, with the balance sheet sufficing for an initial rough analysis. However, detailed identification requires an in-depth analysis using a comprehensive Fixed Asset Register, which may lead to the exclusion of certain assets or simplified valuation based on existing Net Book Values (NBV).

In the professional valuation literature, identification in the context of the valuation of Machinery and Equipment (M&E) primarily refers to the identification of the assets to be valued in the sense of determining or recognizing or characterizing the assets to be valued. A distinction is made between macro-identification and micro-identification. In the context of valuation for financial reporting purposes e.g. a Purchase Price Allocation (PPA) in accordance with International Financial Reporting Standards (IFRS) 3, a detailed sampling and clustering process are essential components of the identification phase, in addition to pure identification.

Sampling becomes necessary when the valuation involves not only a single asset but also multiple or a large number of assets, which typically cannot be individually assessed in detail. This approach is somewhat analogous to portfolio valuation.

If the valuation context or scope necessitates a "full scope" detailed appraisal of all assets, the identification phase may be abbreviated or even omitted. This scenario often arises when the valuation target involves not entire production sites but rather a few high-value individual assets within transactions or accounting realignments. In such cases, the identification phase is brief, with all relevant assets undergoing detailed valuations.

As shown in Figure 4.2, the identification of an asset portfolio is carried out in a top-down manner, while the subsequent valuation is carried out at the level of individual

DOI: 10.1201/9781003622413-4

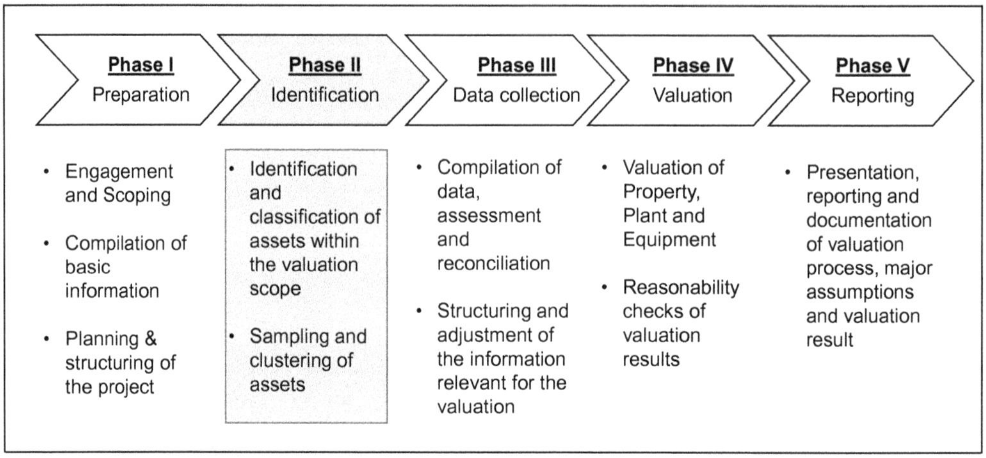

Figure 4.1 Five Phases Approach to Fixed Asset Valuation.
Source: Adapted from Plock et al. [2025], p. 20.

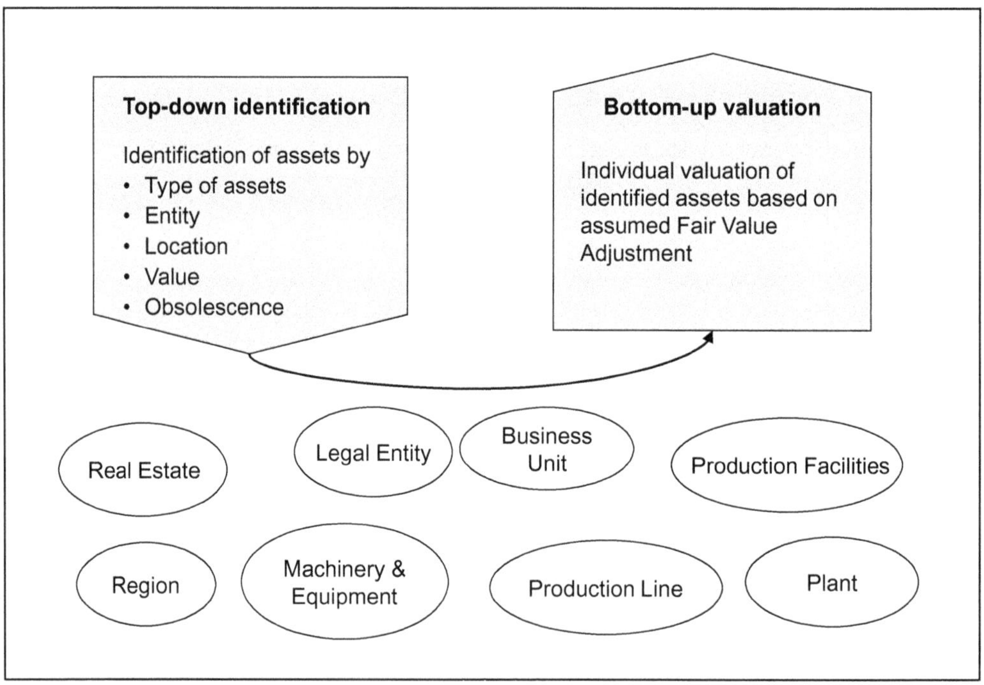

Figure 4.2 Identification of Fixed Assets.
Source: Authors.

assets (bottom-up), and the total value of all assets in scope is calculated as the sum of all individual values of the assets. The identification can be performed on different levels such as legal entity, country/region, location, production line or business unit level. This is typically performed in a structured identification process.

4.1 Identification Process

The process of identifying Fixed Assets originates less from Real Estate valuation than from the valuation of M&E. The simple reason for this is that Real Estate valuation, particularly the valuation of individual properties, involves assets that are relatively obvious to identify and separate.

This is not always the case with M&E assets. This is preliminary due to the fact that the assets involved in M&E valuations are typically significantly smaller and more numerous than those involved in Real Estate valuations.

The identification process for fixed asset valuation involves identifying the key assets to be valued and defining the valuation framework. This is primarily based on financial accounting. For an initial rough analysis e.g. as part of an initial fee calculation, this may be the balance sheet. For a detailed identification, which can then be used in the valuation to exclude or simplify assets, or to value them on the basis of their existing book value, an analysis based on a detailed Fixed Asset Register must be carried out.

From a valuation perspective, a distinction is made between macro-identification and micro-identification.[1] Macro-identification is the identification of the type of asset. In the case of production facilities, this involves looking at the type of production facility, what it produces and what key machines and equipment it comprises. Micro-identification, which is more at the plant level, deals with the characteristics of individual machines and systems.[2]

In practice, M&E is often valued at the level of a company or even a group. This means that all the Fixed Assets of a company are valued and therefore identified. The starting point for identification must therefore be at the group level, as shown in Figure 4.3. From there, a top-down approach is taken to identify the assets at the underlying "levels." Below the corporate and, where applicable, business unit level, this is typically the level of the location or production site, the production line, the individual machine itself or even individual plant components.

4.1.1 Fixed Assets Identification at Corporate Level

If the scope of the valuation does not relate to a single machine but covers all of a company's tangible Fixed Assets, identification must be carried out at company level as described earlier. The starting point for the analysis is then usually the financial accounts with the accumulated book values of all asset groups, including Real Estate and M&E. Since land is not depreciated, for the identification exercise the asset group Real Estate is further subdivided into land as well as Buildings and Constructions (B&C).

When the valuation subject involves the Fixed Assets of a large corporation possessing assets across multiple entities, which are located at various sites and potentially spread across different countries, it is common practice not to inspect and evaluate all Fixed Assets in detail. Instead, locations are clustered and structured to ensure an efficient approach.

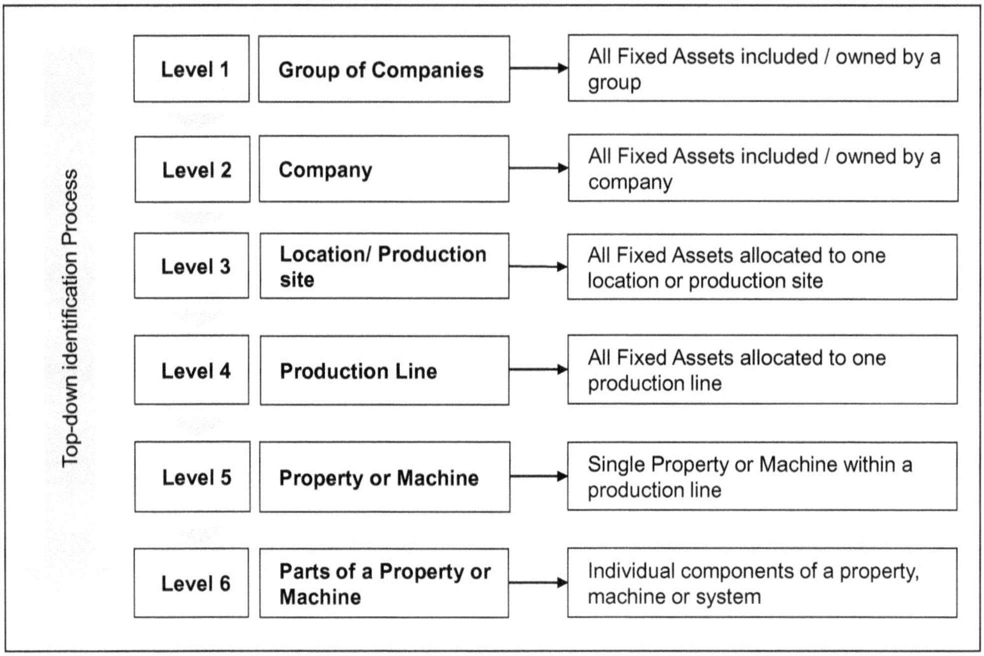

Figure 4.3 Top-Down Identification.
Source: Authors.

In USDm	**No. of**	**Land**		**B&C**		**M&E**		**Total**		**% of**	
Entity	**Plants**	**GBV**	**NBV**	**GBV**	**NBV**	**GBV**	**NBV**	**GBV**	**NBV**	**GBV**	**NBV**
Company A	2	2.5	2.5	9.4	7.2	9.0	7.0	20.9	16.7	16%	28%
Company B	1	1.4	1.4	5.5	3.0	8.4	3.2	15.3	7.6	12%	13%
Company C	5	12.5	12.5	35.6	14.6	45.0	7.5	93.1	34.6	71%	59%
Company D	0	0.0	0.0	0.0	0.0	1.2	0.2	1.2	0.2	1%	0%
Total Group	8	16.4	16.4	50.5	24.8	63.6	17.9	130.5	59.1	100%	100%

Figure 4.4 Identification of Fixed Asset on Company Level.
Source: Authors.

In larger groups, Fixed Assets are often held in separate companies, and not all companies have Fixed Assets. This usually has tax or operational reasons. Therefore, it is important to have an overview of the companies that hold the significant Fixed Assets before starting to identify individual assets. As shown in Figure 4.4, the Gross Book Value (GBV) and NBV of all the Fixed Assets of the group's subsidiaries included in the scope of the valuation are analyzed to determine which subsidiaries own the most Fixed Assets.

In the illustrated example, the majority of tangible assets, both in terms of number of production locations and GBV (87%) and NBV (87%), are concentrated in companies

A (GBV: 16% and NBV 28%) and C (GBV: 71% and NBV 59%). Company B holds a limited number of Fixed Assets (GBV: 12% and NBV 13%), while Company D possesses almost no Fixed Assets (GBV: 1% and NBV less than 1%). Given that Company D owns neither land nor B&C and M&E assets are rather limited, it is likely that the assets categorized under M&E consist solely of office furniture or potentially leasehold improvements in leased premises.

Once the entities containing Fixed Assets have been identified, the next step can commence, focusing on the identification of Fixed Assets from a more qualitative perspective. This involves assessing the nature and condition of the assets, understanding their operational significance and considering any specific characteristics that may impact their valuation. This qualitative analysis complements the initial quantitative assessment based on financial records and helps provide a more comprehensive understanding of the asset portfolio.

4.1.2 Macro-Identification

Macro-identification in the context of M&E valuation, according to the American Society of Appraisers (ASA) pertains to the systematic process of categorizing and comprehending the broader economic and environmental factors that exert influence over the value of Fixed Assets. Fixed Assets, which include property, plant and equipment, are frequently impacted by external factors beyond the immediate control of the organization. Therefore, an accurate valuation necessitates the recognition of these macroeconomic and sectoral influences.

Unlike Real Estate valuation, which is relatively straightforward, the analysis of M&E assets – particularly within a manufacturing or processing environment – can be much more complex. This complexity arises from the interdependencies between the various components of the manufacturing process and the need to understand how these elements contribute to the overall operational capacity of the plant.

The macro-identification phase endeavors to study the entirety of the manufacturing process by identifying the major components that contribute to the plant's design capacity. According to the ASA within this methodology, valuers should address the following three questions[3]:

1 What does the plant produce?

This question necessitates a thorough understanding of the plant's output, including the type of products manufactured and their specifications. Identifying what the plant produces is a fundamental aspect of the valuation process, as it directly correlates with the economic value and market demand for the plant's products.

2 How is the product produced?

This question explores the production process and methodologies employed within the plant. It involves analyzing the technological processes, machinery and equipment used, and assessing their efficiency and effectiveness. A comprehensive understanding of the production process is crucial for evaluating the operational capabilities and potential limitations of the plant.

3 What is the capacity of the plant?

Determining the plant's capacity entails evaluating its design capacity and operational throughput. This involves assessing the maximum output the plant can achieve under optimal conditions, which is a critical factor in understanding the plant's economic value and its potential for scaling production to meet market demands. It also includes the understanding if there are any potential capacity extensions.

By addressing these questions, the macro-identification phase ensures a comprehensive valuation of the Fixed Assets within the broader context of their operational and economic environment. The valuer also needs to get an understanding of how the plant and its capacity fit into the market and compared to competitors' plants.

4.1.3 Micro-Identification

While in the context of macro-identification, the focus is on the location and production processes, whereas micro-identification dives deeper into analyzing individual assets and their components. The ASA outlines several characteristics that facilitate the identification of machinery during the micro-identification process.[4] This involves pinpointing the specific features of equipment and compiling a detailed inventory of each machine. Critical elements include the brand name, model number, serial number, power type, dimensions (when feasible), as well as the age and condition of the equipment.

Brand Name/Manufacturer: The brand name or manufacturer's name is often the sole identifier available. In some cases, identification may be impossible if the nameplate is missing or illegible.

Model Number and Size: While the model number may coincide with size, it often represents the manufacturer's cataloging identifier without directly relating to size.

Serial Number: This can be a vital piece of information, enabling contact with the manufacturer to ascertain the year of manufacture, capacity, size and occasionally the original owner.

Capacity: Capacity is frequently indicated by a machine's size. For instance, a backhoe tractor-loader's capacity might be gauged by the volume it can excavate, a crane by the tons it can lift and a milling machine by its horsepower or kilowatts and worktable size.

Physical Size: Measuring the physical dimensions or footprint of equipment can be beneficial, such as the length of a conveyor belt, the diameter and height of a tank or the span and travel of a crane.

Age and Condition: The chronological age can often be determined from the manufacturer's plate or serial number. Condition is a subjective evaluation of the machine's physical appearance and sometimes its operational performance.[5]

Not all valuation projects allow for the individual identification of all machinery based on the above criteria due to the sheer volume of assets involved. Consequently, identification is typically performed for a sample of assets. A practical approach is to identify a selection of the largest machines or to focus on the key machinery within a particular asset category.

A common scenario where large quantities of Fixed Assets, particularly M&E, require comprehensive identification is during valuations conducted for financial reporting purposes.

4.1.4 Identification for Financial Reporting Purposes

As part of the valuation for financial reporting purposes e.g. PPA in accordance with IFRS 3, all Fixed Assets acquired as part of the transaction need to be identified and valued, and the NBV of each individual asset needs to be replaced by its Fair Value. The focus is generally on those assets for which the difference between the carrying amount at the balance sheet date and the expected Fair Value is the greatest.

This Fair Value Adjustment, also referred to as a "step up" if positive and a "step down" if negative, can have a significant impact on the financial reporting of the entity that owns the assets.

In practice, the phenomenon can be observed that the actual usage and hence useful life of Fixed Assets, especially for B&C and robust M&E with relatively low technological complexity, is often longer than the useful life assumed in the financial accounts for its depreciation.

This means that these assets can actually be used for a longer period than is shown in the financial accounts. As the valuation is not based on the accounting useful life but on the actual economic useful life and the corresponding remaining useful life of the asset, if this is longer than the accounting useful life, the Fair Value will tend to be higher than the net carrying amount.

Therefore, the identification analysis focuses not only on the absolute amount of assets, but it is also necessary to analyze the depreciation of Fixed Assets in each enterprise. Therefore, the ratio between GBV and NBV i.e. the level of depreciation for accounting purposes of all enterprises in the scope, is analyzed during the identification phase for financial reporting purposes (see example in Figure 4.5 below). The companies with the highest depreciation i.e. where the assets are very old but still in use, are also expected to have the highest Fair Value adjustments.

In the illustrated example in Figure 4.5, the depreciation of Companies B, C and D is particularly high compared to Company A. It can therefore be assumed that the difference between the NBV and the potential Fair Value of those Fixed Assets is particularly significant here. However, it should be noted that the book value of Company D's Fixed Assets represents only about 1% of the group's total Fixed Assets. Therefore, although the Fair Value Adjustment may be large in relative terms, the total Fixed Assets of the company are relatively small in absolute terms compared to the group as a whole, making Company D rather insignificant.

Company A's Fixed Assets are relatively new and have not been depreciated much. Therefore, it can be assumed that the difference between the Fair Value and the book value might not be too great. However, due to the absolute amount of Fixed Assets in the company, even small percentage increases in value can have a large absolute impact.

As shown in Figure 4.6, the identification analysis based on the financial accounts can be performed not only at the asset class level, but also at the sub-class level to provide a more granular picture of the individual allocation of Fixed Assets within companies or even locations.

Asset classes for which the Fair Value Adjustment is expected to be relatively and absolutely small are classified as "not material" and do not need to be valued separately. This concept is also referred to as the materiality threshold. In these cases for assets which fall under the materiality threshold, the NBV of the assets as of valuation date is considered to be a sufficient indicator of their Fair Value and hence the NBV is used as the Fair Value.

In USDm				Depreciation		Potential
Entity	**Asset Group**	**GBV**	**NBV**	**in m**	**in %**	**FV Adj.**
	Land	2.5	2.5	0.0	0%	low
Company A	Building & Constructions	9.4	7.2	2.2	23%	low
	Machinery & Equipment	9.0	7.0	2.0	22%	low
	Subtotal Fixed Assets	*20.9*	*16.7*	*4.2*	*20%*	*low*
	Land	1.4	1.4	0.0	0%	low
Company B	Building & Constructions	5.5	3.0	2.5	45%	high
	Machinery & Equipment	8.4	3.2	5.2	62%	high
	Subtotal Fixed Assets	*15.3*	*7.6*	*7.7*	*50%*	*medium*
	Land	12.5	12.5	0.0	0%	high
Company C	Building & Constructions	35.6	14.6	21.0	59%	high
	Machinery & Equipment	45.0	7.5	37.5	83%	high
	Subtotal Fixed Assets	*93.1*	*34.6*	*58.5*	*63%*	*high*
	Land	0.0	0.0	0.0	NA	NA
Company D	Building & Constructions	0.0	0.0	0.0	NA	NA
	Machinery & Equipment	1.2	0.2	1.0	83%	high
	Subtotal Fixed Assets	*1.2*	*0.2*	*1.0*	*83%*	*not material*
Total Group		130.5	59.1	71.4	55%	

Figure 4.5 Identification of Fixed Asset for Financial Reporting Purposes.
Source: Authors.

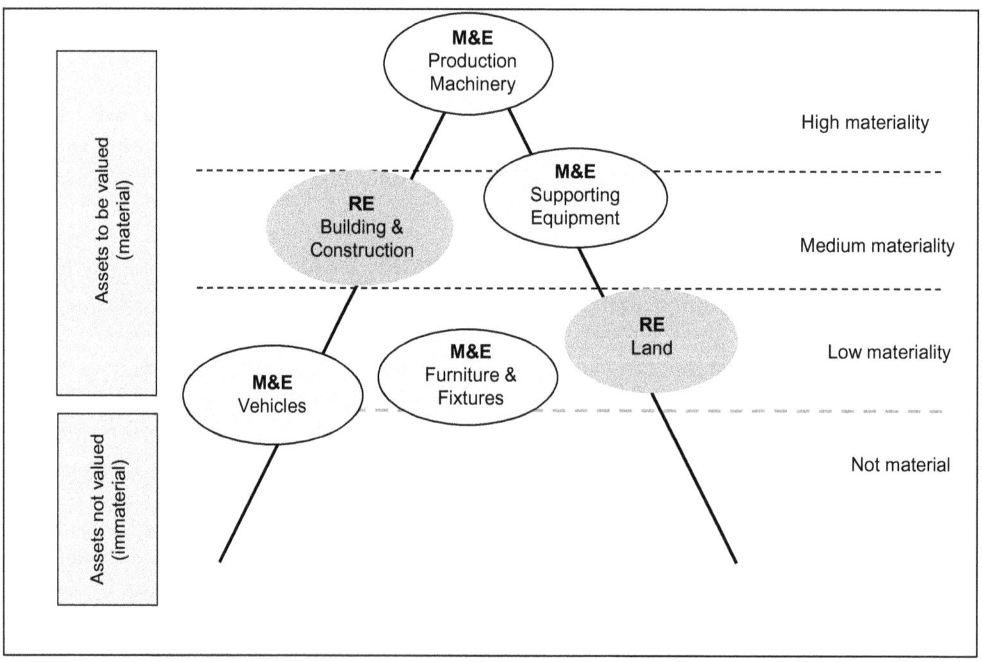

Figure 4.6 Identification of Asset Sub-Groups.
Source: Authors.

In determining the materiality threshold, consideration should be given not only to the impact at the level of the individual asset, but also to the potential impact on the aggregate of all assets that fall below that threshold. It is important to agree this threshold in advance with both the client and other project participants, in particular the company's auditor. If the threshold is not acceptable to the auditor, it may be necessary to lower the materiality threshold and include additional assets in the scope of the valuation.

In most cases, the valuation will not involve a detailed analysis of a single machine or plant, but rather an assessment of large plant portfolios, as described above, of all assets belonging to a group. Since, particularly in large valuation projects involving multiple locations, it is not possible to inspect and value every plant or machine individually for reasons of time and resources, the valuation must be approached using a sampling method.

4.2 Sampling and Clustering

Sampling and clustering are essential parts of the identification process when valuing large asset portfolios and are carried out at the beginning of a valuation process. Sampling refers to the selection of a sample of assets that is as representative as possible (and particularly reasonable) and is then analyzed in detail as part of the valuation process. The results of this analysis are used to draw conclusions about the population of assets to be valued.

Clustering refers to the grouping of the Fixed Assets portfolio in order to focus the valuation on the most valuable companies, production facilities or assets.

4.2.1 Sampling

Sampling is a method employed in Fixed Assets valuation to estimate the value of extensive quantities of Fixed Assets without necessitating the inspection of each individual item. The sampling procedure may involve a certain percent of the total assets value or valuating several assets within each equipment category. Sampling is typically suitable only when it is possible to accurately extrapolate the characteristics of the entire set from the sample.[6]

A key limitation of sampling is that valuers do not observe every individual asset and hence have to make assumptions about the items that remain unexamined. Prior to implementing the sampling approach, it is important for the valuer to communicate the constraints of this valuation method to the client, secure the client's consent to employ the technique and explicitly disclose the approach in the valuation report.[7]

In the context of a valuation for financial reporting e.g. a PPA, the focus will usually be on the asset with the largest difference between the carrying value at the reporting date and the expected Fair Value.

If, because of the reason for the valuation or the scope of the valuation, all assets need to be valued in detail as part of a full scope valuation, the identification phase may be very short or may not be necessary. In practice, this will be the case if the valuation is performed for certain statutory reasons which require an inspection or even stock take of all assets.[8] In practice, however, this is rarely the case for most valuations and the identification phase requires the subject assets to be clustered.

4.2.2 Clustering of Fixed Assets

In valuation practice, it has proved useful to divide the Fixed Assets of different companies or (production) sites into different clusters. Different characteristics may be decisive for the assignment to a particular cluster. Typically, clustering is determined by the following criteria of the assets:

- Legal entity
- Type of assets
- Location
- Value
- Obsolescence
- Age or depreciation of the assets
- Other criteria

In individual cases, it may be necessary to give special consideration to a particular asset class or location in the valuation, even if it would not fall into the highest category according to the above criteria.

In the context of a valuation project to prepare Fixed Assets for financial reporting, close consultation with the auditor takes place during the identification phase regarding the grouping and the sample to be valued. Only if the auditor agrees with the approach does it make sense to start the valuation. If the auditor only sees the approach after the entire valuation has been completed and does not agree with it, the entire valuation may have to be restarted, or at least additional assets included in the valuation.

As shown in Figure 4.7, in practice three different clusters can usually be assumed: e.g. Cluster A, Cluster B and Cluster C.

Cluster A contains by far the largest part of the Fixed Assets but does not necessarily have to include many production sites. Industrial companies often have a historically grown headquarters where a large part of all Fixed Assets are already tied up due to the company's history. In practice, a kind of 80/20 rule often applies i.e. 80% of the total value often located at 20% of the production sites. These 20% of locations must be included in Cluster A, as a small number of locations or production sites already cover a large proportion of the assets to be valued. Cluster B comprises smaller or newer locations where a medium to low amount of Fair Value and adjustment is expected.

Cluster C comprises locations where there is generally no significant production or only small production machinery. When valuing property, plant and equipment in larger structures, this cluster usually includes sales offices that do not have significant M&E, but mostly only office furniture and fixtures and tenant improvements, if any.

The Figure 4.8 shows a typical allocation / structure of cluster (the percentages given are for guidance only). Depending on the asset structure, they should be approximately in the range of Cluster A (70–80%), Cluster B (15–20%) and Cluster C (5–10%). For smaller companies, a full take may be appropriate.

It may also be the case that only two clusters are divided e.g. with a head office in Germany (70%) and a production site abroad (30%), only A and B are structured. Or all assets are valued in full scope.

For Cluster A, detailed information is needed for the valuation while for Cluster C more general information and/or information resulting out of analysis from Cluster A and B are sufficient.

In USDm				Depreciation		Potential	
Entity	**Asset Group**	**GBV**	**NBV**	**in m**	**in %**	**FVAdj.**	**Cluster**
Company A	Land	2.5	2.5	0.0	0%	low	Cluster C
	B&S	9.4	7.2	2.2	23%	low	
	M&E	9.0	7.0	2.0	22%	low	
	Subtotal	*20.9*	*16.7*	*4.2*	*20%*	*low*	
Company B	Land	1.4	1.4	0.0	0%	low	Cluster B
	B&S	5.5	3.0	2.5	45%	medium	
	M&E	8.4	3.2	5.2	62%	medium	
	Subtotal	*15.3*	*7.6*	*7.7*	*50%*	*medium*	
Company C	Land	12.5	12.5	0.0	0%	high	Cluster A
	B&S	35.6	14.6	21.0	59%	high	
	M&E	45.0	7.5	37.5	83%	high	
	Subtotal	*93.1*	*34.6*	*58.5*	*63%*	*high*	
Company D	Land	0.0	0.0	0.0	N/A	N/A	Cluster C
	B&S	0.0	0.0	0.0	N/A	N/A	
	M&E	1.2	0.2	1.0	83%	high	
	Subtotal	*1.2*	*0.2*	*1.0*	*83%*	*not material*	
Total Group		130.5	59.1	71.4	55%		

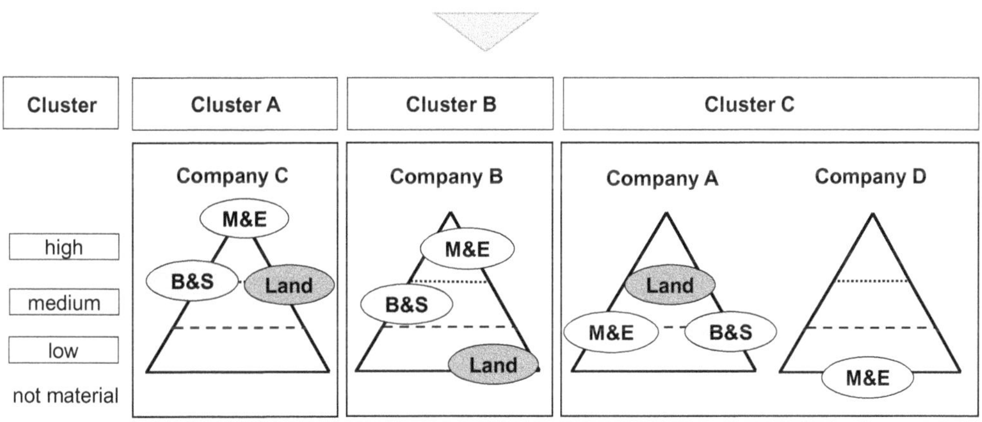

Figure 4.7 Identification of Fixed Asset for Financial Report Purposes.
Source: Authors.

The valuation approach for each cluster is described below.

Cluster A

A full-scope valuation involves a complete valuation of the assets and a physical inspection of the facilities. During the on-site inspection, the physical existence of the assets can be verified and documented based on a sample (e.g. top 10). The inspection provides an impression of the condition and nature of the assets. In addition, if possible, meetings with the technical management of the production site should be arranged and held

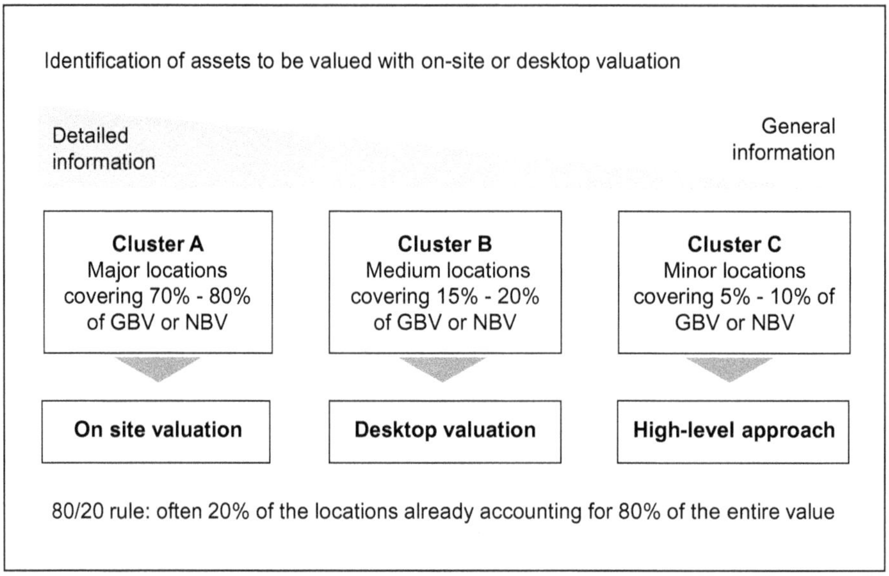

Figure 4.8 Identification and Clustering of Fixed Asset for Financial Report Purposes.
Source: Authors.

during the inspection. This can provide important information for the assessment (see data collection).

Cluster B

The limited scope valuation, also known as a desktop analysis, involves an assessment of the assets without an on-site inspection. The scope of the valuation must be decided on a case-by-case basis. In most cases, the valuation is carried out in exactly the same way as for Cluster A, except that there is no on-site inspection. This can be replaced by video conferencing or photos provided by the client. In individual cases where Cluster B is already significantly lower than Cluster A in terms of the value of the assets, simplified valuation methods may be used. However, this must be possible within the scope of the present valuation and agreed with the client.

Cluster C

Based on the analysis and valuation results for clusters A and B, a rough valuation of the assets in Cluster C is performed. This can be done e.g. by using benchmarks derived from the evaluation of clusters A and B, or by performing a regression analysis of the key figures from the determined GBV, NBV and Fair Values of the assets.

It is important in clustering analysis that all significant assets are assessed with sufficient accuracy. How this can be done in individual cases must be determined separately for each project. Hence, the result of the identification phase and the further valuation should be discussed and agreed with the client, if necessary, with the auditor. For this purpose, so-called identification meetings are usually held.

Notes

1 Iannacito [1989], p. 27.
2 Also see ASA [2011], pp. 32–35.
3 ASA [2011], p. 33.
4 Iannacito [1989], pp. 18–26.
5 Cf. ASA [2011], p. 33.
6 Cf. ASA [2011], p. 37.
7 Cf. ASA [2011], p. 37.
8 See e.g. Statutory Valuations according to Chinese Valuation Standards [2013].

5 Data Collection, Validation and Reconciliation

A robust data foundation is essential for a reliable valuation, as the quality of available data is directly and causally linked to the quality of the valuation result. Consequently, data gathering and the quality assurance process of the data quantity structure constitute a critical step within the overall valuation process. In general valuation literature, which predominantly addresses theoretical aspects, this component of valuation is often treated as comparatively less significant and less discussed.[1]

However, this perspective does not accurately reflect valuation practice, where actually the situation is quite the opposite. Since valuers should be well-versed in theoretical valuation frameworks and valuation procedures and given that valuations are frequently executed using standardized financial valuation models, the practical challenges and time demands, particularly in standard valuations, are often not theoretical but rather focused on establishing a well-founded quantitative structure and data foundation as the basis for valuation.

Therefore the following chapter focuses on the process of gathering, validating and analyzing valuation relevant information and data.

As shown in Figure 5.1, the data collection phase starts after the finalization of the identification and is the pre-requirement for Phase Four, the valuation.

This activity requires knowledge of traditional market and data research techniques but also deep technical, economic, legal understanding of the specific industry and often also specialized accounting knowledge.

5.1 Data Request and Information Required

The data gathering, research and analyses relates on the one hand to property or asset data i.e. information on value-determining features, functions and specifications of the valuation object, and on the other hand to market information i.e. macro- and microeconomic key figures on market activity.[2]

While the asset data is usually provided to the valuer by the client or gathered on site during inspection and/or management interviews, the market data must be researched and determined by the valuer himself (Figure 5.2). Moreover, market data may be further differentiated between market and comparable data. Market data relate to the overall economic market and industry development and comparable data are KPIs (Key Performance Indicators) for the specific valuation assumption.[3]

These two categories differ in that the asset data is usually of a comparatively objective nature, whereas the market information to be researched or the valuer's market assessment would contain a certain subjective element.

DOI: 10.1201/9781003622413-5

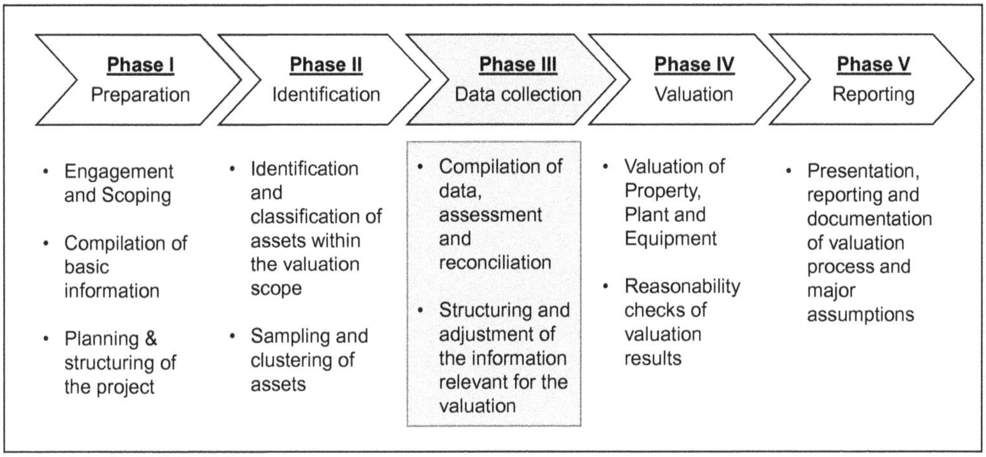

Figure 5.1 Five Phases Approach to Fixed Asset Valuation.
Source: Adapted from Plock et al. [2025], p. 20.

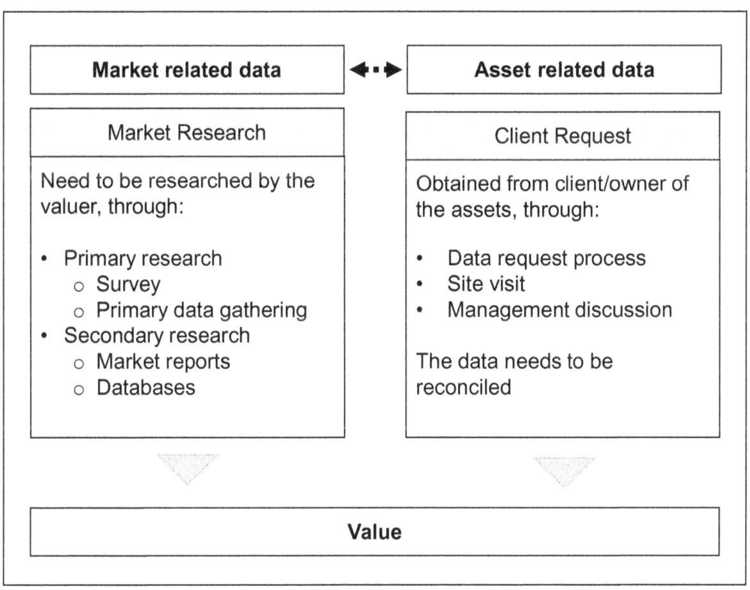

Figure 5.2 Market Data vs. Asset Data.
Source: Authors.

For example, the dimensions of a plot of land or the square meters of rental space of a building can be precisely determined by measurement and mathematical calculation. The same applies to the capacity or volume of a machine. In this context, it is generally possible to obtain fairly precise information, which can then be used as a basis or input for valuation. It is the same with machines and systems.

While the capacity of the machine is given, the corresponding market information on comparative prices is often not absolute. The asset data is therefore mostly of an objective nature, while the market data is subject to subjective assessments by the valuer. An exemption might be the applicable total- or respective remaining useful life of an asset, which is often a judgmental parameter rather than an exact mathematical one.

However, this does not mean that obtaining asset data research and analyses is a simple task. On the contrary, in the case of large asset portfolios with a waste number of individual assets, the data collection, structuring and validation of the asset's information and thus its value-influencing characteristics can take just as much time as their value determination.

In the following it will be distinguished between Market and Asset data collection and due to the different approach a further distinction will be made between Real Estate and Machinery and Equipment.

5.5.1 Data Collection in Relation to the Valuation Scope

The scope of a valuation project, whether full or limited, initially does not affect the general data request to a client. Even if a reduced valuation scope is commissioned and agreed in the engagement letter, all valuation-relevant data for the object will typically be requested in the first step.

There are instances where, due to special circumstances, an indicative assessment of the asset's value is needed in a short period. In such cases, detailed information might be omitted, and simplification assumptions are made for the valuation. However, this approach should be clearly laid out and well-justified in the valuation documentation and reporting.

Sometimes it can be the other way round, the data availability dictates the scope, or limitation in scope may arise from limited data availability. For instance, in a planned company acquisitions and so-called "outside-in" valuation analysis, the target company might not be informed of the takeover intentions by the potential buyer, and hence only publicly available information is accessible for valuing the assets. In such cases, the valuer relies on that publicly available information, such as financial statements and the company's internet presence. From those sources valuation-relevant details are often available only in aggregated form, restricting the valuer to a highly aggregated valuation approach, which can lead to significant variability in valuation results.

For Real Estate valuations, key parameters like location, usage type, and even approximate area are relatively easy to somewhat reasonably estimate. However, information on lease agreements, which can significantly influence the value, is generally unavailable in "outside-in" evaluations. This is similarly true for Machinery and Equipment, where estimating the quantitative scope of the valuation object is challenging or even not possible. Without further corporate details, quantities can only be guessed from publicly accessible financial data. Information about the type, acquisition cost, average age, and book value of assets can provide a basis for rough value estimates although not available on individual level but aggregated level only.

It becomes evident that limited data availability often restricts the scope, not the other way around. In such cases, where indicative values are more estimated based on limited data rather than thoroughly valuated, these limitations should be clearly outlined in the report (see Procedure Disclaimer).[4]

5.1.2 Market Information

When obtaining market data, a distinction must be made between movable and immovable assets. Due to the inherent local nature of Real Estate, its relevant market is significantly more local than for movable assets. This has a serious impact not only on the supply and demand situation when assessing the value derivation but also when researching data. Moreover, the research needs to be separated already due to the availability of date. The data source for Real Estate and for Machinery and Equipment is rather different. While there are typically local market reports with market data and KPIs available often even for different types of property usage classes, this is not the case for Machinery and Equipment.

5.1.2.1 Real Estate-Related Market Information

Immobility and longevity of Real Estate make the location of a property a crucial factor in its timely valuation.[5] All relevant market data are linked to that location and hence drive the entire research exercise.[6]

When assessing a location, the characteristics are typically determined using a catalog of location factors, which, as illustrated in Figure 5.3, is divided into macro- and micro location factors,[7] as well as hard and soft location factors.[8] Their individual importance also depends on the type, usage and characteristics of the individual property.[9]

The macro location encompasses the broader interconnected region of metropolitan areas or cities. In contrast, the micro location provides a detailed view of the immediate surroundings of the property being valued.

Hard location factors are physical elements primarily focused on determining the spatial extent and structure of a location. These include parcel location and property composition, surrounding developments, infrastructural accessibility, and proximity to local service centers. Soft location factors are socioeconomic and psychological elements, where the location's image plays a significant role.[10] A typical soft location factor is prestige, influenced by the quality of construction and the attractiveness of the environment or "address."[11]

Both soft and hard location factors in micro and macro locations play a crucial role in Real Estate valuations. However, their weighting varies significantly depending on the asset class or usage: residential, office, or industrial properties, due to their specific influences on a property's value and attractiveness.

For residential properties, the region's economic stability and growth potential are essential. Factors include general economic conditions, unemployment rates, and demographic trends. A growing job market and positive population development can increase housing demand. On the micro scale, proximity to schools, shopping centers, public transport, and recreational facilities is vital. Soft factors like quality of life, social environment, and neighborhood safety also play a major role, significantly influencing residential property attractiveness and price trends.

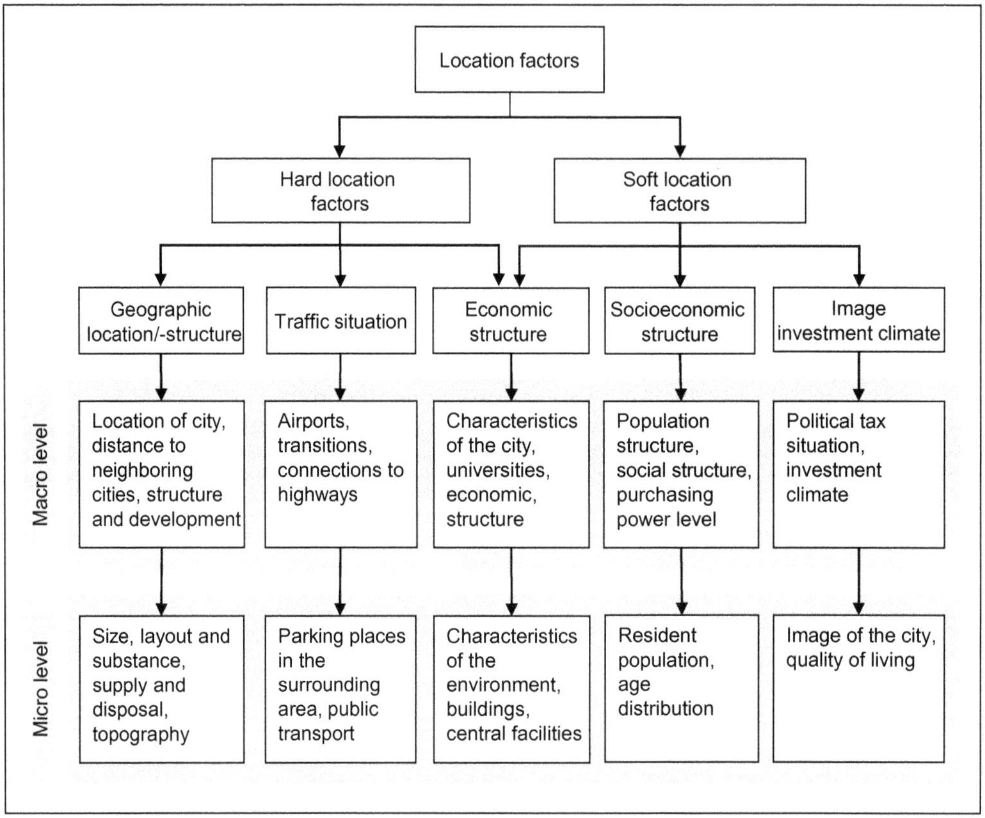

Figure 5.3 Location Factors.
Source: Adapted from Gondring et al. [2024], p. 249.

For office properties, regional economic dynamics are critical. A strong business presence and good connectivity to national and international transport networks are hard factors enhancing location attractiveness. Availability of skilled labor and proximity to other business hubs are key considerations. On the micro level, access via public transport, parking availability, and proximity to restaurants and services are important. Soft factors like location image and architectural quality of surroundings also impact office property attractiveness.

For industrial properties, infrastructure and logistical connectivity are crucial. Hard factors like proximity to highways, ports, or airports and availability of utilities (e.g. electricity, water) are central. Economic policy frameworks and availability of support programs can also influence location choice. On the micro level, the size and expansion potential of the site, proximity to suppliers and customers, and workforce availability are critical. Soft factors like environmental conditions and support from local authorities can also play a role.

Both soft and hard location factors, divided into macro and micro aspects, are vital for Real Estate valuation. Their weighting varies by asset class, with residential properties more influenced by soft factors, while office and industrial properties rely more on hard factors.

Besides common macroeconomic indicators like national and local economic growth, interest rates, unemployment rates, tax conditions, and inflation expectations, typical Real Estate-specific market information and data include[12]:

Real Estate Economic Market Data

- Development of Real Estate transaction amount and volume
- Vacancy rates
- Construction price index series
- Market price index series

Comparable property data

- Land reference values
- Property sales comparisons
- Market comparison rents
- Total useful lives
- Discount rates, capitalization rates
- Gross initial yields and net initial yields
- Management costs, administration fees
- Comparison factors for developed land
- Conversion and adjustment coefficients

The macroeconomic related market information should give the valuer an indication of where in the Real Estate market cycle he/she should "position" the asset as of valuation date and what further market development may look like. Local comparable property data give an indication for the applicable valuation parameter like rental income, outgoings and yield.

While the focus of researched data is somewhat related to the chosen valuation method, often multiple methods are applied for plausibility purposes, striving to gather as much market information and data as possible during market research.

5.1.2.2 Machinery and Equipment-Related Market Data

In the market analysis of Machinery and Equipment, the research approach fundamentally differs from that of Real Estate valuation. While Real Estate is primarily dependent on local markets due to its fixed location, Machinery and Equipment are mobile and can be evaluated in a national or even international context. This necessitates a different approach and the consideration of specific factors. Since most Machinery and Equipment are not bound to a fixed location, they can be deployed in various geographic markets. This requires an assessment of international market demand and the competitiveness of the machinery in different regions as well. Factors such as transportation costs, tariffs, and international trade agreements can also play a role.

Hence, unlike Real Estate, for Machinery and Equipment assets the industry in which Machinery and Equipment assets can be utilized is of critical importance. The analysis should focus on the supply and demand and growth potential of the specific industries where the machinery can be employed and the related products which are produced by the Machinery and Equipment assets. This includes technological developments, industry trends, and regulatory frameworks that may influence the market development and demand.

Technological advancement is a crucial factor in valuing Machinery and Equipment. The pace at which new technologies are developed can affect the lifespan and value of existing machinery. Therefore, a market analysis should consider the rate of innovation in the relevant industry and assess how well the machinery aligns with future technological developments.

In addition, a comprehensive competitive analysis is essential to determine the Market Value of Machinery and Equipment. This involves identifying major competitors, their market shares, and analyzing pricing strategies. The competitiveness of the machinery in terms of price, quality, and technological features should also be evaluated. This includes exchange rates, global economic cycles, and international demand trends that can affect Market Value.

Key specific market information regularly includes:

- Price development / indices for Fixed Assets
- Price development / indices for Machinery and Equipment
- Comparative metrics for new prices relative to capacity
- Comparative purchase prices for used Machinery and Equipment
- Prices for new but comparable Machinery and Equipment
- Installation costs
- Maintenance costs
- Transportation costs
- Rental rates for Machinery and Equipment
- Operating costs for Machinery and Equipment

The market analysis of Machinery and Equipment requires a comprehensive examination of the industries in which they can be used, as well as international markets. Unlike Real Estate valuation, which heavily focuses on local location factors, Machinery and Equipment require consideration of global economic and technological trends.

Besides the aforementioned macroeconomic indicators, industry-specific metrics in Machinery and Equipment valuation are significantly more important than in Real Estate. The more industry-specific the equipment, the more likely the specific industry sector will need to be analyzed, as supply and demand dynamics significantly impact the machine's value.

5.1.3 Real Estate Asset Information

While the researched market information and comparable data provides the foundation for the valuation assumptions, the asset information constitutes the quantitative structure of the valuation exercise.

The basic difference of collecting asset related information compared to collecting market data is that the asset information will be obtained from the client or target respectively. This can be done by requesting data and documents from the client as well as discussions with the client and by inspecting the valuation object. In most valuation it is all of the above. A request list is sent to the client, who provides the data which are then verified in a site inspection and discussed in a management meeting.[13]

Some of the information required is the same for Real Estate and Machinery and Equipment, while other aspects differ. To ensure an efficient process, it is advisable to request the necessary data from the client in a structured manner using a comprehensive "request list." Furthermore, the information should be precisely described. In practice, it might be beneficial to review the data request list together with the client to avoid misunderstandings regarding the requested data and ensure a smooth data delivery process.

In smaller projects with a manageable data volume or when there is only one contact person available on the client's side, the data collection is typically not a problem. However, in complex large-scale projects where data needs to be collected from various national subsidiaries and numerous contact persons are involved, it is important to establish a structured data request process and meticulously track the received data and its sufficiency for the valuation. This involves recording when and to whom requests were sent, when and from whom data was received, whether the obtained data is sufficient for the valuation, or if there are follow-up questions regarding the received information necessary, and which information is still pending. The longer the data gathering process takes, the later the valuation can start to be finalized. Therefore, potential delays in the project timeline that could affect adherence to the agreed submission deadline for the valuation project should be communicated to the client early on to avoid surprises at the end of the project.

5.1.3.1 Real Estate-Related Asset Information

Within the data request, all relevant technical, economic and legal asset-related information is requested.[14] The following list contains the essential information, but in individual cases significantly more detailed or more extensive information may be required and requested from the client.

Real Estate

Land

- Location (complete address)
- Ownership status of the land (owned or leased)
- Copies of land registers / title deeds
- Area: Total sq. m. of freehold / leasehold land per property including a breakdown by type of land such as total land, thereof surface land, thereof industrial, agricultural/wood/greenland, residential, other
- Copy of official site plan / cadastral map
- Lease contract or heritable building right contract (if existing)
- Description of location (especially the infrastructure)
- Status of planning laws and building regulations (Zoning Plan / Local Building Plan)
- Status of actual structural use (floor space index, site occupancy index, cubic index)

Building and Infrastructure

- Area: Square meters, (optional cubic meters) per freehold buildings per property split by usage (e.g. production, logistics, R&D, office, locker rooms, labs, warehouses, technical and central area, other such as guest houses, residential)
- Year of construction per building plus year of major renovation/ modernization

- Site layouts / ground plans on which all buildings (and floors) are identifiable
- Parking places (outdoor parking lots, underground parking, carports, etc.)
- Description of the outside facilities (paved exterior surfaces, walls and fences, entrance gates, etc.)
- Information regarding major repairs, renovation or redevelopment measures
- Information regarding the protection of historical monuments and other legal and official restrictions
- Contracts about hereditary building rights, if applicable
- Operational costs on a yearly basis subdivided by types of costs
- If applicable: Overview of any (sub-)rental contracts (rent roll as a data file)
- Major rental contracts (if applicable)

Real Estate (leasehold)
Leasehold contract including:

- Name of tenant
- Type of use of rental space
- Area of rental space
- Monthly/annual lease payments rent increase clauses
  - Turnover rent
  - Index-linked rent
  - Graduated rent
- Lease duration
- Agreed payment of operating cost
- Extension option(s)
- Other value-relevant information
- Any additional agreements regarding the leased object

Building Improvements (installments in leased buildings)
This includes owned improvements and installations in leased buildings.
List of all leasehold improvements including the following information:

- Entity # and Entity description
- Location code / location description
- Business unit/area allocation (e.g. building #)
- Asset Class # and Asset Class Description / Name
- Cost Center # and Cost Center Description / Name
- Asset ID # and Sub Asset ID #
- Asset Description or Name of Asset
- Historic Date of Acquisition and Date in Service
- Historic Cost of Acquisition / Gross Book Value ("GBV"), including received allowances, donations if existent – please make sure to provide the true historic cost when the item was first acquired new

For reasons of simplification, in some cases not all of the above listed data is requested in an original detailed format but in a more aggregated format. This e.g. applies to tenant lists, so-called rent rolls which are typically requested in a processed and aggregated format. If all tenancy agreements for a large number of tenants e.g. in a large office building or even in a shopping center, have to be analyzed on a rental contract level and the information relevant to the valuation extracted, this is a very time-consuming task for the valuer. Therefore, in practice, a rent roll is often requested directly from the client. This document is maintained by the asset management of the owner and should include all pertinent information regarding individual lease agreements.[15]

Even if the client confirms the accuracy of the information in the rent roll, the valuer should at least carry out a sample review. For this purpose, the rental agreements e.g. for the ten largest leases are requested, and the corresponding data is cross-checked with the rent roll. If different types of lease agreements exist, a sample contract of each type of contract should be included in the sample. This process ensures both efficient procedures while maintaining high data security.

5.1.3.2 Machinery and Equipment Asset Data

In the valuation of Machinery and Equipment, the indirect cost approach and the trending method are commonly employed. This process involves determining the reproduction cost new based on historical acquisition costs and relevant price indices, from which Fair Values are derived, taking into account depreciation. The foundation for this approach is a comprehensive quantity structure of all assets to be valued, typically derived from the client's Fixed Asset Register.

As depicted in Figure 5.4, data collection based on individual invoices and quotations is often impractical for large asset inventories due to the frequent lack of documentation and the significant time investment required, particularly when dealing with extensive asset holdings.

Consequently, in practice, the first option is regularly utilized, wherein the client is requested to provide the relevant Fixed Assets Register. This register should include at least the key information listed below, which is sent to the client as a request list within the context of the valuation. A typical request list which is provided to the client includes the following information requests and questions:

Machinery and Equipment

Fixed Asset Register as of valuation date[16]

- Entity # and Entity description
- Location code / location description
- Business unit/area allocation (e.g. building #)
- Asset Class # and Asset Class Description / Name
- Cost Center # and Cost Center Description / Name
- Asset ID # and Sub Asset ID #
- Asset Description or Name of Asset
- Depreciation life (Total Useful Life) in periods (state if months or years)
- Historic Date of Acquisition and Date in Service
- Historic Cost of Acquisition / Gross Book Value ("GBV"), including received allowances, donations if existent
- Accumulated depreciation
- Net Book Value ("NBV")

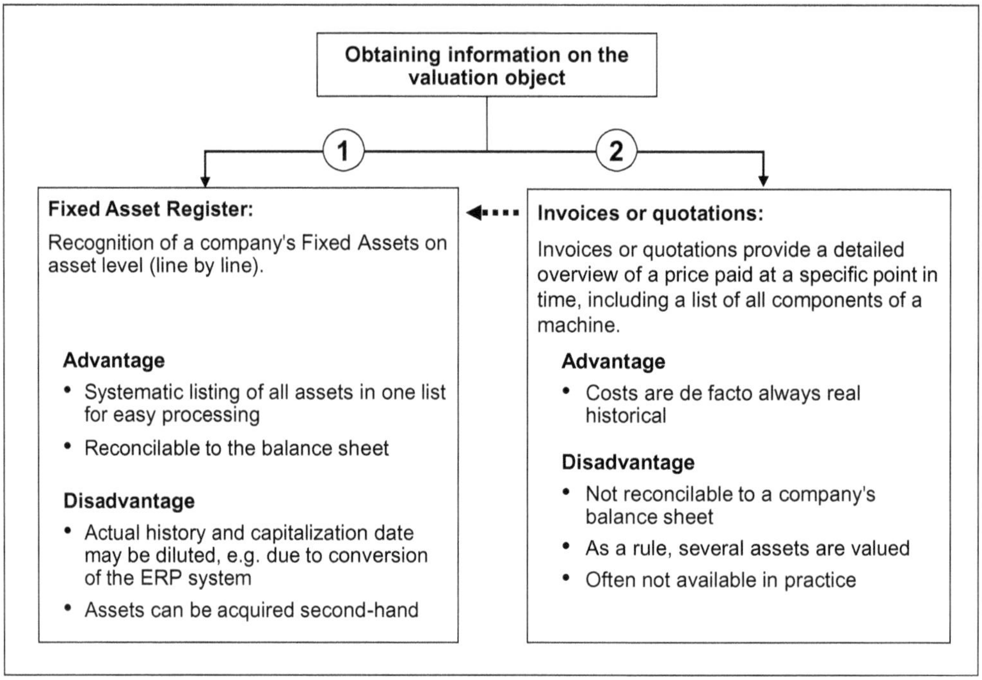

Figure 5.4 Obtaining Information on the Valuation Object.
Source: Adapted from Plock et al. [2025], p. 31.

- Type of depreciation (e.g. straight-line, double-declining etc.)
- Reference to balance sheet item (Land, Building infrastructure, Equipment, Construction in progress, and Right-of-use assets)

Additional questions include:

- Out of Use Assets
- Sold
- Scrapped
- To be scrapped
- Held for sale
- Technically installed capacity by production unit or per plant and key performance metrics e.g. overall equipment efficiency
- Historic utilization/production output, YTD and planned utilization
- Indications for obsolescence like excessive energy cost, extreme maintenance cost, low utilization, technologically outdated equipment, out-of-use equipment, etc.
- Product-specific attachments or tools
- Planned CapEx required split by Preservation CapEx (stay-in-process CapEx), Development CapEx and Strategic CapEx

As shown in the list above, in many cases, the data requirements are quite extensive because the valuer must consider all valuation-relevant information in the assessment.

However, certain data are more crucial than others. For instance, in Real Estate valuations, a detailed construction description is useful but not as critical as the data on area or lease agreements. Similarly, in Machinery and Equipment valuations, maintenance information is beneficial, but the asset register holds greater significance for the valuation process. Therefore, it can be advantageous to prioritize data requests to ensure that the client focuses on providing the most important information. Core data, especially to set up the quantitative framework, is essential to initiate the valuation. Other information might be more relevant later when assessing specific parameters and could potentially be submitted by the client during the valuation process without causing significant delays in the project's timeline. Or certain assumption can be made and then later on validated and potentially fine-tuned with the correct/final data.

Historical Cost of Acquisition

Therefore the historical cost of acquisition is pivotal in Machinery and Equipment valuation, unlike in Real Estate, where available standardized construction costs per property type often allow for a direct cost approach. For Machinery and Equipment, reproduction cost new values are often unavailable since there are no comparable official cost sources available like for buildings and structure, necessitating reliance on historical costs indexed to the valuation date to establish Fair Value. It is crucial that these costs reflect the actual original acquisition costs, not estimates from potential acquisitions or purchases of used machinery. Verification can be achieved through invoice sampling.

When relying on historical cost data from the accounting it is important to understand the relevant accounting standard applied and hence which costs are included in the accounts.

The accounting standard IAS 16 outlines the recognition of initial acquisition and production costs in paragraph 16, which apply to all acquisition or production costs of property, plant, and equipment at the time they are incurred. These costs include those originally incurred for the acquisition or construction of the asset, as well as subsequent costs for expansion, replacement, or maintenance. The valuation upon recognition is conducted at the acquisition or production cost.[17]

According to IAS 16, the cost of acquisition or production of a tangible asset includes:

a) The purchase price, including import duties and non-refundable purchase taxes, after deducting discounts, rebates, and allowances;
b) All directly attributable costs necessary to bring the asset to the location and condition necessary for it to be capable of operating in the manner intended by management;
c) The initial estimated costs of dismantling and removing the item and restoring the site on which it is located.

Furthermore, examples of directly attributable costs[18]:

- Costs of employee benefits arising directly from the construction or acquisition of the item of property, plant, and equipment
- Costs of site preparation
- Initial delivery and handling costs
- Installation and assembly costs
- Costs of testing whether the asset is functioning properly
- Professional fees

When using the historical cost of acquisition for the valuation of Fixed Assets in order to apply the trending method, it has to be ensured, that the cost reflect the following key requirements:

1. **Real Historical**
   The actual historical acquisition cost without changes, such as valuations from previous purchase price allocations.
2. **Discount-Free**
   Market practice dictates that discounts or special conditions should not be considered in the valuation (contrary to IAS 16.16).
3. **Activation of Costs**
   Activation or capitalization of costs should correspond to the time of acquisition.

It is crucial to ensure that costs recorded in financial accounts reflect the actual original acquisition costs, as outlined rather than estimated costs from potential company acquisitions or purchases of used machinery. Furthermore, these costs must represent the original costs at Market Value at the time of acquisition, which can be verified, at least for a sample of relevant assets, by checking invoices.

Going Concern vs. Liquidation

When considering costs in valuation, it is fundamental to distinguish whether the valuation is conducted under the "going concern" assumption, meaning the continuation of the current use at the existing location, or if the asset is to be relocated. The latter scenario typically applies in a Liquidation Value assessment, where costs related to deriving the reproduction new values may not be applicable.

Therefore, during the analysis of historical acquisition costs, it is essential to carefully examine which costs pertain to the actual asset and which represent additional costs. These additional costs may need to be treated as "sunk costs" in the context of the valuation.

By adhering to these principles, appraisers can ensure that valuations are accurate and reliable, providing a clear representation of asset value across diverse portfolios.[19]

5.2 Site Visit and Inspection

The on-site visit and inspection of the valuation object is a fundamental part of each (full scope) valuation. It is particularly important since the valuer can confirm the physical existence of the asset and moreover get a firsthand understanding and impression of its condition.

For Real Estate valuation site visits also give the valuer an opportunity to not only inspect the property but get a better understanding of the surrounding and micro location as well.

5.2.1 Real Estate Site Visit

Numerous aspects that become apparent during a property and site inspection can have an impact on the perception of the value of the property in question on the market. In the valuation of Real Estate, this can include the following aspects, which are defined by the RICS for a site visit for the valuation of land and buildings[20]:

- Characteristics of the surrounding area and availability of communications and facilities that affect value
- Characteristics of the property or investment
- Dimensions and area of the property and buildings
- Construction method of the buildings and their approximate age
- Types of use of the land and buildings
- Condition of the premises, extensions, fittings and technical building equipment
- Apparent structural condition and state of maintenance
- Natural hazards e.g. instability of the ground, risk of flooding from any source e.g. rainfall and rivers
- Hazards that are not of natural origin e.g. contamination, any physical restrictions on further development or construction

In valuation practice, the ability to convert qualitative information like impressions during site visit into quantitative data plays a crucial role. This process, which can be referred to as "quantitative transformation capability," describes the methodical conversion of subjective observations into measurable and comparable values. This transformation is particularly important for systematically incorporating qualitative attributes into the valuation and establishing an objective basis for analysis.

The quantitative transformation capability involves several steps, including the operationalization of qualitative attributes, the development of scales for evaluating these attributes, and the application of benchmarking methods. For instance, the age of a building might be operationalized through a shorter remaining useful life, while the aesthetic appeal of a house could be assessed using a scale to quantify its impact on Market Value.

By applying these methods, qualitative impressions are ensured not to remain merely subjective but are instead transformed into a form that can be utilized in valuation practice. Thus, the quantitative transformation capability is an essential component of valuation work, allowing qualitative information to be converted into a quantitative form that can be used for informed decision-making and analysis by considering related assumptions in the valuation or adjusting the valuation assumptions.

While this process was often individually performed by the appraiser in the past, there are now tools available, particularly in Real Estate valuation, that allow impressions and insights to be systematically captured in electronic data processing (EDP) programs during inspections. These inputs are then automatically integrated into valuation models, providing suggestions for potential impacts on the valuation parameters. Although this does not replace the appraiser's transformation capability, it at least aids in the structured assessment of qualitative experiences on-site. It also helps for a consistent and structured approach, especially when multiple parties are involved in the simultaneous inspection of large Real Estate or asset portfolios.

5.2.2 Inspecting Machinery and Equipment

For Machinery and Equipment, the American Society of Appraisers (ASA) also lists the following procedures and information that can be carried out as part of a site inspection for movable interest-bearing equipment[21]:

- Conduct a comparison between the descriptive information provided and the actual physical item it pertains to.

- Verify that the inventory or stock-keeping unit quantities supplied are reasonable. Note that this verification can only be performed with real-time data.
- Record observations regarding the quality or condition of the inventory.
- Document the stocking method, including references to specific locations.
- Assess the capability to control access to the inventory in the event of a future sale from the present site.
- Evaluate the ability to display and make the inventory accessible for inspection by potential buyers in the event of a future sale from the current location.
- In cases where perpetual records are unavailable or when comparing to a dated physical inventory, efforts must be made to estimate the current quantities on hand. This typically requires assistance from individuals involved in the last physical inventory. If no records exist from which to estimate, the appraiser must make a diligent effort. Estimating value without detailed records via site inspection is extremely challenging and subject to significant variations and should only be used as a last resort.
- For a manufacturer's inventory, note the physical transitions as inventory progresses from raw materials to work-in-progress (WIP) to finished goods.
- Record whether obsolete inventory is physically segregated on site.
- Note inventory that appears inactive, indicated by the presence of dust, rust, or other signs. Further investigation into items exhibiting these characteristics is warranted.
- Identify specialized or proprietary items within the inventory.[22]

Together with a site visit it is common in valuation projects to also organize management meetings with the client and or target / owner of the asset.

5.2.3 Management Meeting

Management meetings play a crucial role within the framework of data gathering and information validation. These meetings provide a platform to ensure clarity, reconcile data, and deepen the understanding of valuation objects and their relevant industry environments. A primary focus during these meetings is the comprehensive handling of the data request list. It is important to eliminate potential misunderstandings through thorough discussions about the details of the request list and received data. This dialogue ensures that all parties involved have a shared understanding of the data requirements, facilitating the efficient collection and validation of necessary information.

The discussion of Real Estate data received thus far is crucial for understanding the buildings subject to valuation. Detailed deliberations may encompass specific Real Estate topics such as legal ownership and leasehold improvements. These aspects are essential for accurately appraising the value of the Real Estate components within the Machinery and Equipment valuation context. Another key component of management meetings involves reconciling data from various sources, including technical specifications, accounting records like the Fixed Asset Register, and market information. This reconciliation process is vital to ensure data consistency and reliability, thereby contributing to an accurate valuation outcome. The meetings also provide an opportunity to request additional data and address queries that may have arisen from the information received thus far. It is essential to gain a deeper understanding of the valuation object, including its characteristics and operational specifics. This process aids in refining the valuation assessment and aligning it with the actual conditions of the Machinery and Equipment. Moreover, these meetings facilitate gaining a better understanding of the market situation within the industry-specific environment in which the company operates. Understanding the detailed production process steps is also essential, as it provides insights into operational intricacies and potential influences on the valuation of Machinery and Equipment.

Detailed questions and procedures discussed in the management meeting are listed by the ASA as following[23]:

- Solicitation of general opinions from company personnel of active or inactive items and obsolete inventory.
- Identification of new products. Confirm their inclusion or exclusion in the inventory records provided.
- Discussion of any future plans that will affect inventory mix (e.g. new products or modification of existing products, entering new markets or discontinuing existing products, changes in ordering or purchasing methods).
- Computer-generated inventory reports that will assist in the valuation obtained through meeting with appropriate parties. Examples include separate reports identifying slow-moving or obsolete inventory, or separate reports for new products or items recently introduced. Where possible, electronic copies compatible with the appraiser's systems also should be obtained to allow for further sorting and analysis of the data.
- Discussions of the company's products and markets, in order to understand the flow of goods and factors affecting marketability.
- Where applicable, discuss any patents, trademarks, licenses, manufacturing rights, etc., that may not be transferable.
- Discuss return privileges (goods returned by customers of the subject company), historical return levels, and methods of disposal used for the returned goods. The company's accounting policy in terms of reserves taken against returns should also be addressed.
- Identify key or primary vendors.
- Discussion with purchasing personnel of the potential for returning materials to vendors. If the inventory appraisal is for a lender, returning goods to the vendor would typically result in the reduction of an outstanding balance, not in cash to the lender. Therefore, returning goods is seldom a viable option in an appraisal for a lender.
- Discuss the company's methods of selling or disposing of slow-moving or obsolete inventory.
- Discuss with purchasing personnel purchases of slow-moving or obsolete inventory or obsolete inventory obtained from other sources and the relative discounting involved.
- Determine if obsolete inventory is included in the inventory data provided. If yes, determine whether any adjustments already have been made to the cost base.
- With regard to proprietary inventory, ask questions relating to possible alternative uses. This would include information for potential buyers in the event the inventory is later liquidated as a result of a product being discontinued.

To properly address those questions, the contact person for Machinery and Equipment related topics is typically someone from the purchase and maintenance department. For Real Estate it is a contact person from the asset management or facility management. In addition, especially to discuss the Fixed Asset Register and the reconciliation of the different accounting data, some contact person from accounting is crucial as well.

By systematically addressing these aspects during management meetings, stakeholders can enhance the accuracy and reliability of Machinery and Equipment valuation, ensuring that all relevant factors are carefully considered in the valuation process.

5.3 Data Structuring and Grouping of Fixed Assets

In the area of fixed asset valuation, particularly concerning Machinery and Equipment, simplification remains a key focus for both appraisers and their clients. While valuations are conducted at the asset level, similar asset groups are often evaluated using uniform valuation approaches to optimize efficiency and consistency. When assessing large portfolios of Fixed Assets, encompassing both Real Estate and Machinery and Equipment, it is advantageous to organize these assets into similar groups. This enables the application of a streamlined yet robust valuation methodology.

The concept of group-level valuation is more prevalent with Machinery and Equipment assets than with Real Estate. For land, grouping is typically inapplicable due to significant value variations influenced by macro and micro-location factors. Conversely, in portfolio valuations of buildings, group-level evaluations are utilized. Typically, most fixed asset valuations are conducted on an individual basis. However, for Machinery and Equipment valuations, individual assessments are less common. Although valuations are performed at the "line item" or individual asset level, the assumptions used in these evaluations are applied across entire asset groups.

A primary challenge in valuing large asset portfolios lies in segmenting the portfolio into highly homogeneous groups. This structuring facilitates several critical objectives. For Machinery and Equipment, homogeneity allows for the selection of similar valuation approaches across the entire group, ensuring consistency and efficiency. For land and buildings, it ensures the highest possible representativeness of sample objects within each group. This is particularly crucial in property portfolio valuations, where insights obtained from a representative site visit are extrapolated to the entire portfolio.

Furthermore, systems of machines and equipment are grouped and evaluated collectively to streamline the valuation process. Complex production processes are dissected into their components to facilitate precise valuation, while the client's cost centers may offer a logical structure for grouping assets. The scope of the engagement dictates the level of detail required in the valuation process. While simplification is essential for efficiency, it must not compromise the necessary detail and accuracy of the valuation.

By adhering to these principles, appraisers can ensure that the valuation process for Fixed Assets is both efficient and thorough, providing reliable and representative assessments of asset value across diverse portfolios.

5.3.1 Definitions of Condition

The condition of the assets is one of the major drivers of its value. There are a large number of definitional descriptions of the condition of assets, ranging from very good to scrap condition. A good classification of the ASA is given below. It is also helpful that the ASA divides the condition groups into five categories based on a Likert scale. This helps in statistical evaluations with metric characteristic requirements to show a class width that is as equal as possible.[24]

Very Good This term refers to an item in outstanding condition, fully capable of performing its intended function without requiring modifications, repairs, or extraordinary maintenance at the time of inspection or in the foreseeable future.

Good Condition This term applies to items that have undergone modifications or repairs and are operating at or near their specified capacity. However, due to the effects of age and/or usage, minor repairs may become necessary, or the item may need to be operated at a slightly reduced capacity compared to its full specification in the foreseeable future.

Fair Condition This term describes items being utilized below their fully specified capacity due to age and/or application effects, necessitating general repairs and the replacement of minor components in the foreseeable future to restore their utilization level to or near their original specifications.

Poor Condition This term is used for items that can only be utilized significantly below their specified capacity, where full capability cannot be realized in their current state without substantial repairs and/or the replacement of major components in the very near future.

Scrap Condition This term characterizes items that are no longer serviceable and cannot be utilized to any practical extent, regardless of the repairs or modifications that might be applied. This condition is applicable to items that have been used for the entirety of their useful life or are completely technologically or functionally obsolete.[25]

5.3.2 Structure of Fixed Asset Register

The Fixed Asset Register regularly forms the basis for the Machinery and Equipment valuation. For Real Estate valuation, the Fixed Asset Register is important as a reference to check the completeness, but due to the size of the assets and nature of Real Estate valuation not quite so crucial.

In most cases, the Fixed Asset Register is provided in an MS Excel format. Even though the Fixed Asset Register is derived from the company's management accounts and is therefore usually an indirect part of the financial reporting, this does not mean that the valuer can accept the data without checking it. The data must be analyzed and validated prior to valuation.

In practice, it is often the case that data is found to be implausible. This can affect both the allocation of assets to the relevant asset classes and the allocation of depreciation.

5.4 Data Classification

Asset groups can be further divided into various asset classes, with the most common being Real Estate (Immovable Assets) and Machinery and Equipment (Movable Assets). Real Estate encompasses land and buildings. Land is categorized into freehold and leasehold land, while buildings are divided into freehold buildings and installations in leased buildings, also known as tenant improvements. Freehold buildings can be further classified based on usage into categories such as office, production, warehouse, residential, and other commercial types.

Machinery and Equipment, on the other hand, are classified based on the useful life of the assets. This category includes plant and machinery, which typically consists of machinery with long useful lives, as well as equipment, which encompasses furniture and fixtures, office equipment, and computer equipment. Vehicles are also considered under this category and are often subdivided into cars, trucks, and forklifts. It is important to note that the grouping of Real Estate and Machinery and Equipment can vary significantly based on the industry. For valuation purposes, it is crucial to group assets into similar categories to ensure accurate assessments.

Fixed Asset Register

Client:

Data as of the valuation date:

1	2	3	4	5	6	7	8
Asset class	Name of the class	ID no.	Location	Cost Center	Description of the object	Acquisition date	Year of construction
						mm/dd/yy	yyyy
9	10	11	12		13		14
Acquisition/ production costs (Gross Book Value)	Carrying amount (Net Book value)	Accumulated amortization	Depreciation method		Useful life		Assignment to Production stages lines or levels
Currency	Currency	Currency	yy	hours	yy	hours	Name

Figure 5.5 Structure of a Fixed Asset Register.
Source: Authors.

In the context of the valuation of Machinery and Equipment assets, structuring these into uniform groups is essential. This approach is necessary because valuation assumptions apply to groups of assets rather than individual items, due to the large number of assets typically involved. According to the ASA, asset groups are described through accounts and classes. A standardized grouping of appraisal accounts and classes ensures uniformity and consistency. For appraisal purposes, accounts are major groupings of similar assets. The basic separation of tangible assets into accounts includes land, buildings (or structures), and equipment, which encompasses machinery and furniture, though some might classify furniture as a separate account. The equipment account can be further divided into various classes such as production machinery, general plant equipment, and office furniture and fixtures.

Utilizing the client's accounts and classes is often preferable, especially when the appraiser's findings must align with the client's accounting records. Although the client's equipment classes may differ from the standard descriptions, they are likely to be similar. This alignment is crucial for maintaining consistency and accuracy in asset valuation.

There are assets that resist clear categorization into specific asset groups, such as cleanroom installations used in specialized production processes within the semiconductor, pharmaceutical, or food industries. These assets, often integrated into the building infrastructure, may be classified either as part of the building or as Machinery and Equipment. It is imperative during the valuation process to ensure that these assets are neither overlooked nor doubly counted. The latter could occur if cleanrooms are attributed to Real Estate under the income approach by using comparable rents for cleanroom spaces, while simultaneously being evaluated as Machinery and Equipment based on acquisition costs and a trending approach. Precise categorization is thus vital to achieve distinct and accurate valuation outcomes.

A further group or category of asset not shown in the Figure 5.6 is assets classified as work in progress. Those assets could be from each subgroup listed above and are assets which have not been activated in the accounting but are still under construction or are in the progress of installation. They are classified in the accounts under work in progress and since they are not fully installed or in use as of valuation date, the accounting depreciation has not started.[26]

5.5 Data Validation

5.5.1 Plausibility Checks, Sample Comparable, Invoices

In principle, a declaration of completeness is requested from the client for valuations. However, this does not release the valuer from checking the plausibility of the data received, as well as checking whether the information received can be correct at all. There are limitations to this. In the case of Real Estate e.g. buildings do not usually have to be re-measured, and no inventory is normally taken for Machinery and Equipment. The valuer must therefore be able to rely to a certain extent on the information received from the client, while at the same time carrying out certain audit procedures or at least plausible checks of the information, depending on the valuation background.

These checks are often carried out through sample examinations.

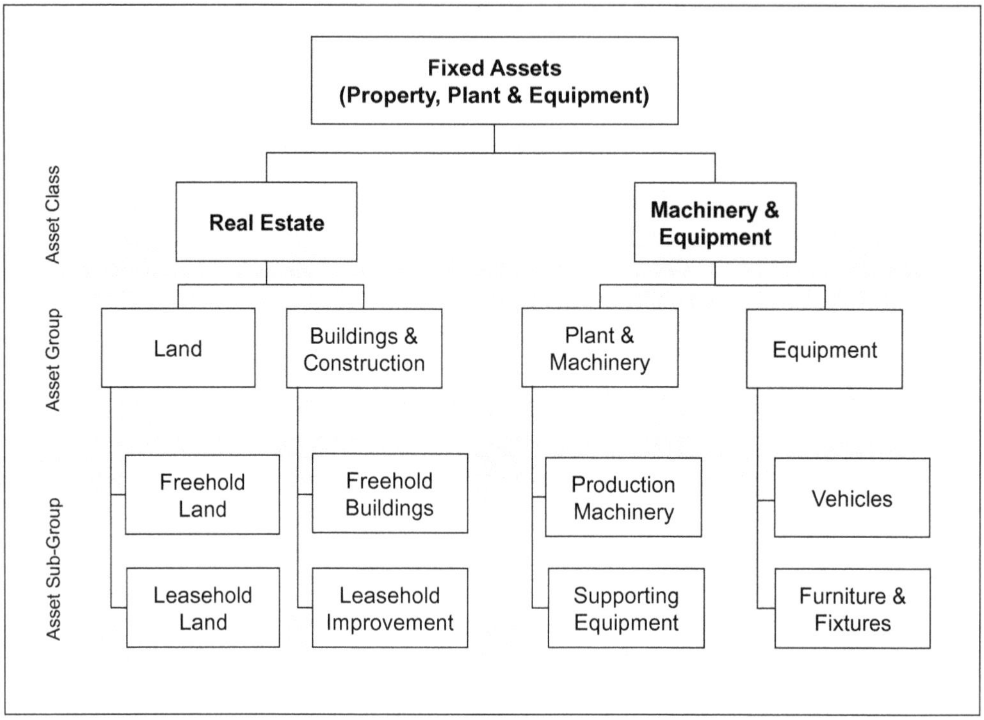

Figure 5.6 Fixed Asset Groups, Classes and Categories.
Source: Authors.

5.5.2 Data Reconciliation from Different Sources

A major challenge in many valuation projects is the reconciliation of information from different sources. As mentioned above, information on the valuation objects essentially comes from three areas:

- Accounting – Information from accounting
- Market – Information from the market
- Technology – Information on the individual assets

The accounting area is particularly important for valuations prepared for financial reporting purposes, as the results of the valuation must be reconciled with the structure of the company's accounting systems. Accounting data also plays a greater role, particularly in the case of Machinery and Equipment valuations, as the Fixed Asset Register regularly forms the database for the valuation.

As part of data preparation, all information from the various areas must be brought together and structured in such a way that they fit together as a result (Figure 5.7).

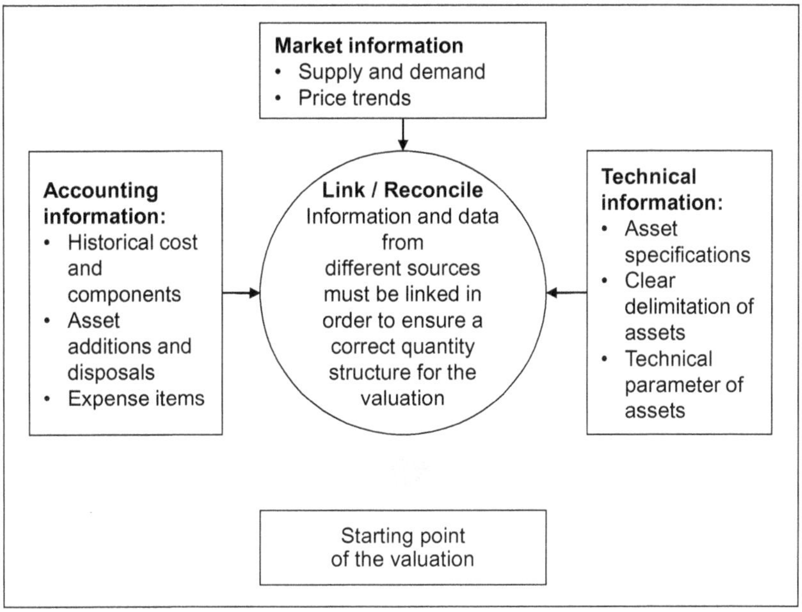

Figure 5.7 Data Reconciliation.
Source: Adapted from Plock et al. [2025], p. 21.

5.5.3 Reconciliation of Fixed Asset Register

To ensure the completeness of the assets being valued, an inventory could, in principle, be conducted as part of the project, documenting all individual assets. However, since in practice this can be too time-consuming for many valuations, the Fixed Asset Register is commonly used as the basis for valuing Fixed Assets.[27] If the scope of the valuation includes all the assets owned by a company or a company's entity, to ensure that the Fixed Asset Register indeed includes all assets owned by the company or subject to valuation, it must be reconciled with the corresponding balance sheet provided by the company. If the total Gross Book Value and Net Book Value in the Fixed Asset Register match those in the audited balance sheet, it can be assumed with reasonable certainty that the Fixed Asset Register is complete.

If there are differences between the Gross Book Values and Net Book Values in the Fixed Asset Register and the corresponding totals in the balance sheet, they need to be resolved. Such differences may arise e.g. if the Fixed Asset Register has been prepared by management in accordance with local GAAP and the company prepares its consolidated accounts in accordance with IFRS, where adjustments may have been made. There may then be differences between the book values of the received Fixed Assets register and the audited balance sheet at group level. If the consolidated balance sheet is the only balance sheet audited by the auditor, the valuer will need to understand the adjustments to ensure that the Fixed Asset Register provided to him includes all the assets belonging to the company and can be used as a basis for valuation. This must then be discussed in detail with representatives of the accounting department so that the differences can be explained.

The Fixed Asset Register can only be used as a complete valuation basis if it can be clearly reconciled with the audited balance sheet. If significant discrepancies cannot be resolved, a physical inventory may need to be carried out as part of the valuation to ensure that all assets within the scope are included in the valuation.

5.6 Ageing Analysis

Whenever the valuation involves not just individual tangible assets but large portfolios of assets, the age structure of the assets to be assessed becomes a relevant consideration. This is particularly important when valuing an entire production site with all its associated Machinery and Equipment. To address this, an Ageing Analysis must be conducted. This analysis is one of the initial steps in the valuation process and can provide crucial preliminary insights even before a detailed data analysis is carried out.[28]

The Fixed Asset Register serves as the foundation for the Ageing Analysis. This analysis examines not only the age or the acquisition date (usually the year of acquisition) but also the relationship between Acquisition Cost and Net Book Value.
The following information is essential:

- Acquisition Date
- Acquisition Costs
- Net Book Value

These details can be found in the Fixed Asset Register. During the Ageing Analysis, the asset register should be filtered by acquisition years and location grouping assets according to their acquisition date. For all assets acquired in a given year, their respective acquisition costs and associated Net Book Values are aggregated. The result is an overview, illustrated in the following Figure 5.8.

The graphical presentation enables a clear structure which a range of information can be easily read from the accounting investment structure at a glance. It is important to point out that the results of the analysis can only be conjecture in qualitative terms, which must then be substantiated in the further course of the valuation.

The aging analysis of a production site shown in the example below allows you to read the following information. It is important not to understand this aging analysis with regard to the GBV and NBV over time, but to consider the respective book value difference of the individual years.

- The production site was probably built/established 34 years ago. The first assets were acquired here, most of which were plots of land that were not depreciated
- The first machinery was purchased 33 years ago (see GBV 33, 32 and 31 years ago). All plants have already been largely or almost completely written off. This can be seen from the high difference between cumulative GBV and NBV. However, these depreciated assets are all still in use. Otherwise, they would no longer be in the Fixed Asset Register.
- Buildings from 31 to 33 years ago were also depreciated. The remaining NBV is very likely representing the building purchased about 30 years ago.
- Furthermore, it can be assumed that an expansion of the site or at least new machinery was purchased around 20 years ago, 10 years ago and 3 years ago. If necessary, an indication of the industry-specific production cycle can already be derived here.

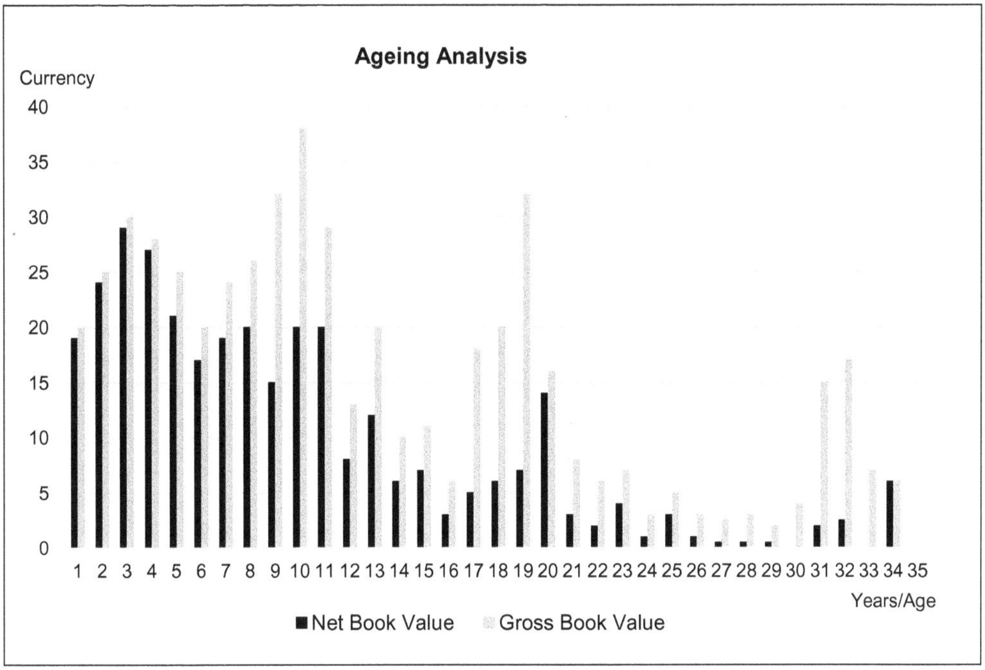

Figure 5.8 Ageing Analysis.
Source: Authors.

- Overall, it can be seen that many assets are heavily depreciated but are still in use. This indicates that the actual total economic useful life of the assets is longer than their accounting useful life. As a result, a comparatively high Fair Value Adjustment (Fair Value vs. Net Book Value) is to be expected here.

Particularly in the context of accounting valuation e.g. purchase price allocations in accordance with IFRS3, it is examined during the Identification Phase in which assets are expected to have a large difference between book value and Fair Value (Fair Value step-up or Fair Value Adjustment). This is particularly the case for assets that are already largely depreciated in accounting terms i.e. have a low or no book value (Residual Value) but continue to contribute to the production and therefore the success of the company. The economic useful life here is obviously longer than the accounting useful life (economic useful life > accounting life). These assets must therefore also be assigned a Fair Value as part of the valuation.

The more assets that have already been largely depreciated but are still in use, the higher the difference between NBV and FV will be. Aging analysis is a good way of analyzing and graphically depicting this relationship, and in practice aging analyses can be carried out at different levels or for different groups of assets. As a rule, this is done at site level in order to determine at an early stage of the project which sites will be valued, and which may not have a material Fair Value Adjustment.

However, it is also possible that an aging analysis is carried out at business unit or regional level. Sometimes it also makes sense to carry out an aging analysis for different investment groups. In this case, the focus of the valuation can be placed on the investment groups that account for the absolute major part of the value but also the absolute largest part of the expected Fair Value Adjustment.

Notes

1 It is rather prominently discussed in the valuation guidelines. E.g.: "The quality of the valuation will, in part, rely on the quality of information [...]." Cf. TEGOVA [2022], p. 55.
2 Also see Reisbeck and Söhne [2009], p. 60.
3 Cf. ASA [2021], p. 51.
4 At this point, however, it should be pointed out once again that a limited scope is not permissible in many cases or for many valuation reasons, particularly in the context of valuations influenced by regulation. All relevant data must be available in order to perform a full scope valuation.
5 Cf. Gärtner [1996], p. 69; Cf. Petersen [2005], p. 188; Fierz further states: "The location determines the divisor according to which land and buildings are divided into the value of a property. The reason for this simple rule is that the yield that can be achieved with the property depends on the location, whereas the construction costs are only insignificantly determined by the location" (Fierz [2001], p. 28).
6 Reisbeck and Söhne [2009], p. 61.
7 Reisbeck and Söhne [2009], p. 61.
8 Cf. Gondring et al. [2004], p. 249.
9 Preuß and Schöne [2016], p. 133.
10 Cf. Isenhöfer and Väth [2000], p. 350.
11 Cf. Gondring et al. [2004], p. 261.
12 Also see Reisbeck and Söhne [2009], p. 66.
13 Reisbeck and Söhne [2009], p. 60.
14 Cf. Reisbeck and Söhne [2009], pp. 62–65.
15 Also see Huber [2019], pp. 200ff.
16 Although the FAR is mainly used for the valuation of machinery and equipment assets, it also contains Real Estate-related information
17 Cf. IAS 16.15.
18 Cf. IAS 16, Paragraph 17.
19 It may also be that individual machines and systems cannot be moved at all or only with great difficulty. This is the case e.g. with mining equipment in mines. These may only be used at the intended location.
20 Cf. RICS [2014].
21 Cf. ASA [2011], pp. 291–292.
22 Cf. ASA [2011], pp. 291–292.
23 Cf. ASA [2011], pp. 292–293.
24 Cf. ASA [2021], pp. 49–50.
25 Cf. ASA [2021], p. 50.
26 Please refer to Chapter 6 for further information of how to value WIP.
27 For certain valuations e.g. statutory valuation inventory procedures to validate the actual physical existence and status of the assets might be compulsory. This needs to be checked with the respective valuation background and related procedures required.

28 In the case of projects in which the assets of a larger company with multiple production locations are valued, the Fair Value Adjustments assumed are to be expected in particular for old locations. Newer locations e.g. in emerging markets, are often only slightly depreciated and therefore often have a comparatively lower FVA. However, this primarily relates to machinery and equipment. Real Estate, especially land, has often risen sharply in value in these locations in particular, as these emerging or frontier markets have often been characterized by very dynamic growth in recent decades.

6 Valuation

The following chapter is about the Valuation Phase, which represents Phase IV within the valuation project process (Figure 6.1). In this phase, the valuation is carried out based on the previous theoretical framework and the data and information obtained so far. The valuation phase represents the central step within the entire project.

This chapter provides a detailed overview of the fundamental valuation concepts, approaches, and methodologies applied for the valuation of Fixed Assets. Therefore, the chapter is structured around three primary valuation approaches: the Market Approach, Income Approach, and Cost Approach.

Moreover, the hierarchy and the choice of valuation approach and method is examined, acknowledging the theoretical equality of the three approaches while noting practical differences and its implications for the valuation practice. The chapter starts with an introduction to valuation concepts.

6.1 Valuation Concepts

Even if the terms approach and methodology are sometimes used synonymously in valuation theory to describe valuation concepts, procedures and methods, they must be kept typologically separate and arranged hierarchically.[1]

As shown in Figure 6.2, there is a hierarchy of terms with regard to valuation. Based on the framework, which can be the applicable international or local valuation law or valuation regulations, there are various approaches to solving the problem of valuing an asset. Under each approach, there are different valuation methods. These are implemented using a valuation model and valuation assumptions parameters.

6.1.1 Framework

The framework of a valuation refers to the applicable conditions, which are primarily determined by the reason for the valuation. Depending on the reason for the valuation and how the client intends to use the valuation report, a number of regulatory requirements may apply. As explained in Chapter 2, certain rules and regulations have to be considered for valuations carried out for external valuation purposes, such as tax or financial accounting purposes, and they can significantly impact the scope, content and procedures of the entire valuation project. These can relate to the approach, methodology, and even valuation parameters of the model.

DOI: 10.1201/9781003622413-6

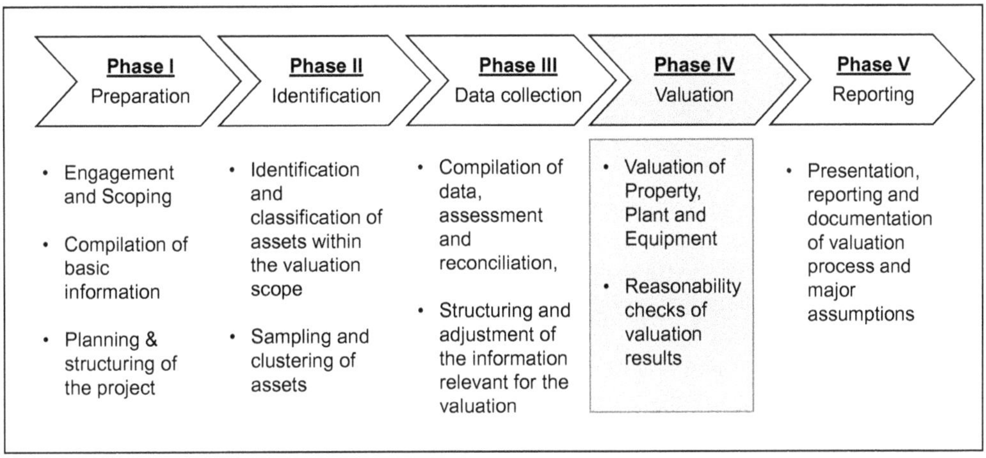

Figure 6.1 Five Phases Approach to Fixed Asset Valuation.
Source: Adapted from Plock et al. [2025], p. 20.

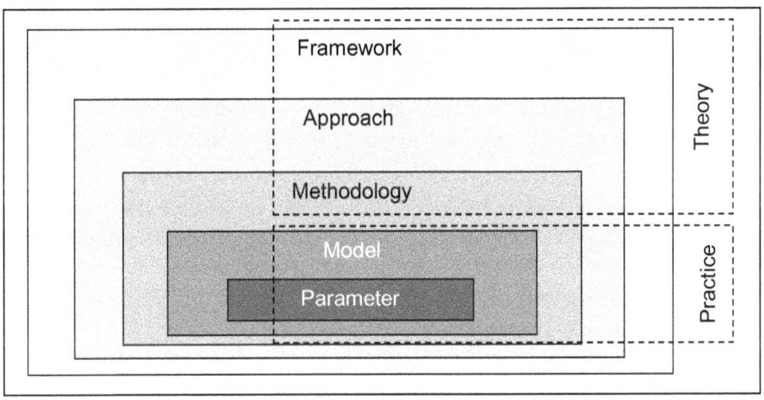

Figure 6.2 Valuation Concepts.
Source: Authors.
Note: Cf. TEGOVA [2022], p. 64, Sehab [2025], online.

Therefore, the appropriate framework must be clarified upfront, as all other basic principles and framework conditions for the subsequent valuation process are based on that framework and its potential regulatory specifications.

6.1.2 Approach

An approach is the overarching concept or idea for solving a problem.[2] In the context of the valuation, the approach is essentially the theoretical basis for the valuation method. Hence, TEGOVA defines an approach with regards to fixed asset valuation as: "The way in which, having regard to the available evidence, the valuer considers how to determine the value of the subject asset."[3]

In line with the general valuation theory, there are three fundamentally different approaches for the valuation of Fixed Assets. These are the[4]:

- Market Approach
- Income Approach
- Cost Approach

Despite their fundamental theoretical differences in their conception all approaches essentially follow the principle that in a free market, supply and demand determine the price of an asset. Each of the three approaches attempts to make comparisons with market developments using different reference points or price indicators to derive the value.

This market reference can be achieved through comparative transactions (market approach), comparing income and the related risk (income approach) or through comparable cost calculation (cost approach). Based on the idea of the respective approach, there are several different valuation methods which can be applied for fixed asset valuation.

6.1.3 Methodology

The methodology or method is hierarchically below the approach and can be described as a structured mathematical process for determining value using the respective approach. TEGOVA defines methodology as: "The particular procedure, based on one or more valuation approaches, used by the valuer to arrive at a determination of value."[5] Each of the above three approaches includes various valuation methods that can be applied using different models. Certain valuation methods are even based on various approaches e.g. the Residual Value method which can consist of elements from the market, income and cost approach.

While the approach is globally the same in different territories, the respective methods can differ in their form and structure. This can be for instance due to different local conditions which can have a significant influence on the design of established valuation methods. Some valuation methods are mathematically rather simple while others can require more complex mathematical calculation models.

Their usage in practice is defined by the valuation background as well as the nature of the asset to be valued and availability of related market data.

6.1.4 Model

The (financial valuation) model is a tool for the practical implementation of the methodology. There are various types of models. Some are more qualitative; others are highly quantitative.[6] Some models can be very simple calculation aids or others can be highly sophisticated and complex financial mathematical calculation models. How simple or complex a model is, depends not only on the approach and the method applied, but also on the respective valuation object and valuation background. Individual occasions require a more precise value derivation, while others allow a more general approach and the valuation model for a vacant parcel of land will likely be less complex than the one of a shopping center. The complexity of the model is usually accompanied by the required granularity of the applied valuation parameters.

Since all market parameters are often projections that must be set by the appraiser, Damodaran[7] points out that the most complex and supposedly most accurate valuation models are not always the ones that most accurately reflect market developments when deriving value. Because they are always based on valuation assumptions that have to be estimated.[8]

To avoid "over-engineering," one could say that the model should be as simple as possible but as complex as necessary to cater for the valuation requirements and reflect the valuation subject's character. Since valuation practice usually involves computer-aided and standardized models anyway, the underlying parameters and approaches are what matter in practice.

6.2 Parameter

For the input parameters used in a valuation model a distinction is made between property or object parameters and market parameters.

Property parameters include floor space and capacity, and in the case of rental properties, details of the rental agreements, such as the amount of rent, the terms of the agreement and renewal options. The client usually provides this information on the property parameters (see Chapter 5).

Market parameters include information relevant to the valuation from the market environment, such as market comparison prices, standard market rents or yield ratios. These must be researched by the valuer.

While property data is almost always objective and forms the basis of the valuation model, applied market information often includes a certain subjective element and must be applied appropriately by the valuer.

The choice of approach, method and subsequently the applied valuation model must always be made in each individual case against the background of the:

1. reason for the valuation,
2. the nature of the valuation object, and
3. the availability of reliable market data.

This choice must be made at a very early stage of the valuation process and has substantial influence on the entire valuation project.

6.3 Valuation Approaches

As described at the beginning and shown in the following Figure 6.3, there are basically three different approaches for the valuation of Fixed Assets: the market approach, the income approach and the cost approach, which each comprises several methods.

The **Market Approach** is a valuation technique in which the value is estimated based on market prices from actual transactions and on asking prices for assets currently available for sale. The valuation process is essentially that of comparison and correlation between the subject asset and other similar assets. Considerations such as location, time of sale, physical characteristics and conditions of sale are analyzed for comparable assets and are adjusted to indicate the current value of the subject asset.

The **Income Approach** derives the value based on the future income that an asset can be expected to generate over its useful life. This approach is typically estimated through the Investment Method, Term and Reversion Method or Discounted Cash Flow (DCF) Method. Basically, under all those methods, the valuation comprises the present value of

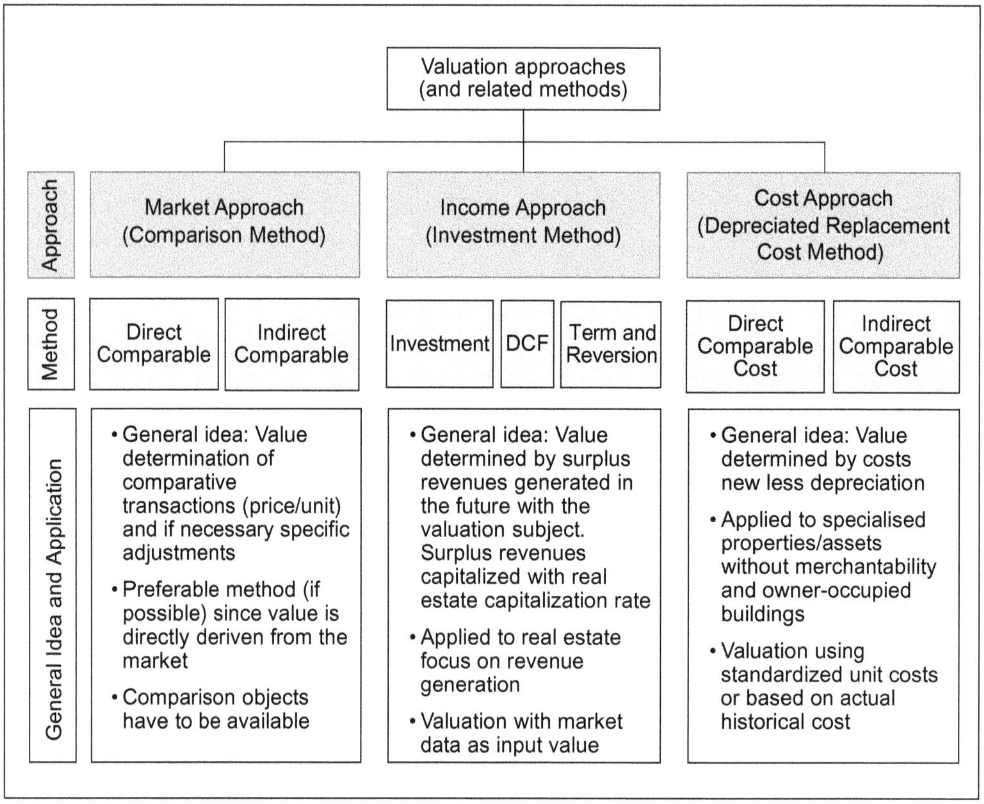

Figure 6.3 Valuation Approaches and Related Methods.

Source: Authors.

Notes: Cf. Hackelberg and Hennig [2021], p. 327; Cf. Leopoldsberger et al. [2016], p. 435.

estimated future cash flows discounted at a rate of return that considers the relative risk of achieving the cash flows and the time value of money.

The **Cost Approach** is a valuation approach that uses the concept of depreciated cost new as an indicator of value. The premise of the Cost Approach is that a prudent investor would pay no more for an asset than the amount for which the asset could be replaced. Replacement cost new (RCN), which refers to the cost to replace the asset with like utility using current material and labor rates, establishes the highest amount a prudent investor would pay. To the extent that an existing asset will provide less utility than a new one, the value of that asset is less. Accordingly, RCN is adjusted for loss in value due to physical deterioration, functional/technical obsolescence, and economic obsolescence.

Due to the generally highly specialized characteristics of Fixed Assets, and often the limited related market activities especially for Machinery and Equipment assets, the cost approach, unlike for other asset classes, takes a dominant role in practice.[9] Therefore, the cost approach and its related valuation methodologies is given particular importance in this chapter and the procedure is discussed in more detail.

6.4 Hierarchy of Valuation Approaches

In theory, all three of the above-mentioned valuation approaches are on an equal footing and, when used correctly, their respective valuation methods should lead to the same result. Nevertheless, the difference between theory and practice is greater in practice than in theory. And, when applying different valuation methods to value the same valuation object in practice the result can differ quite significantly.[10] And thus, in the literature as well as related accounting frameworks, a hierarchy of valuation approaches is regularly mentioned.[11]

The International Financial Reporting Standards (IFRS) provide a "Fair Value hierarchy" in its Standard IFRS 13 Fair Value Measurement.[12] IFRS 13 regulates the valuation of assets and liabilities as well as the valuation assumptions and methods to be used for this purpose. IFRS 13 requires the most objective measurement possible. The input parameters for Fair Value measurement should be used in a defined order of priority, three levels.

The first level includes assets for which quoted, unadjusted prices for identical assets can be determined on active markets. The second level includes assets and liabilities that cannot be classified in Level 1 but can be derived from directly or indirectly observable prices. These include valuations based on quoted prices for similar assets or liabilities in active markets. Level 3 includes assets and liabilities whose valuation includes input factors that are not based on observable market data.[13]

Level 3 will be relevant for most Fixed Assets that are rarely traded due to their design and special nature and for which there are usually no sufficient comparable transactions available on the market.

In line with IFRS 13, in literature, the market approach is often described as the best valuation method, as the value is derived directly from observable market developments. The result therefore reflects the aggregated value assessment of buyers and sellers.

The second is the income approach, which is also closely linked to market developments through its major input factors income and yield and reflect investors' potential future financial benefits of the assets.[14]

The cost approach is often regarded as less preferable. It is argued that the cost of an asset does not necessarily reflect the price an investor would be willing to pay for it, since he would be less interested in what the asset has cost than what the future benefits are.

Moreover, the cost development lags somewhat behind the trends in market price developments. Thus, the cost approach is referred to in the literature even as a "method of last resort"[15] if the other two approaches cannot be applied. However, this does not mean that the cost approach is not a suitable valuation method for Fixed Assets, one just must understand and consider the potential limitations caused by the possible decoupling of cost orientation from market orientation.

Due to the high degree of specialization, especially of Machinery and Equipment assets, it will often be difficult or even impossible to find sufficient comparable transactions on the market. Comparables are regularly observable for standardized assets e.g. standard production machines, vehicles or computers. But when it comes to more specialized assets such as specially constructed foundations for heavy machinery or special support structures on buildings for crane systems this will be more difficult.

In principle, the more specialized the production system or machine is, the more difficult it becomes to find sufficient comparable transactions. This is particularly the case with assets specially manufactured for a plant, production line or even with specialized assets manufactured by the producing company itself. Often, no comparable transaction will be found on the market for these, since there practically are none. Moreover, individual research for all assets as part of the valuation would be too time-consuming and resource intensive.

Therefore, in practice, the market approach will play a very minor role, at least as a valuation method for Machinery and Equipment assets, with the exception of "standardized" very marketable assets such as cars, forklifts or computers. However, it is often used for the purposes of verifying the plausibility of results derived by other approaches/methods.[16] Where on the other hand it is often used is for the valuation of undeveloped land or, in individual cases, very marketable buildings.

6.5 Choice of Valuation Approach and Method

In order to choose a valuation approach and subsequently the most appropriate method to value the subject asset on hand, a structured decision process should be applied as outlined in Figure 6.4. This structure ensures a well-founded and transparent decision for an approach and methodology and should be properly documented in the valuation report.

Following the "hierarchy of value" and the related preference for the application of the market approach, in a first step it needs to be assessed if comparable market transactions are available for the valuation subject. While for local Real Estate markets a sufficient number of comparable property transactions (Land and Buildings) is often available at least for standard usage types of properties, as outlined above for Machinery and Equipment assets, this is often not the case. However, it needs to be highlighted that the lack of market activity and available market data needs to be explained and outlined in the valuation report. Practical obstacles such as a limited professional budget of the valuer or a

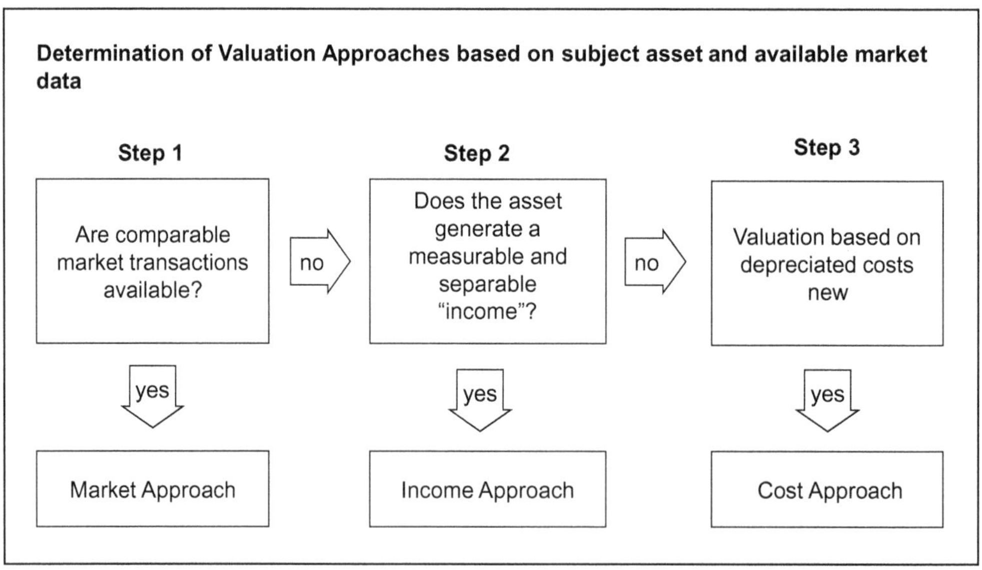

Figure 6.4 Determination of Valuation Approach.
Source: Adapted from Plock et al [2025], p. 10.
Note: In accordance with IFRS13, Paragraph 72–90.

tight time schedule for the valuation process cannot be accepted for reducing efforts for sufficient market research. Only if after a proper market research and investigation the valuer can prove that no reasonable data are available and thus the market approach cannot be applied sufficiently, the first question "are comparable market data available" can be answered negatively and step two can be applied.

As shown in Figure 6.4, if there are not enough available market transactions, and hence the market approach is not an option as the preliminary valuation approach, the question to be answered is, whether a "measurable and separable income stream" can be identified for the individual asset, since this would be the basis for the second choice: the income approach.

Machinery and Equipment assets can have very different sizes and functions. What they usually have in common is that they work together, particularly in the context of production facilities and create cash flows jointly together, e.g. a production line or overall entire production facility. It is therefore difficult, if often impossible, to break down the revenue of a company or production site meaningful to individual assets according to their individual share in the overall revenue generation. A separate cash flow might be possible at the level of the entire plant or production lines, but not actually at the level of the individual tangible assets. In practice the question arises: which part of a company's cash flow should be allocated to an individual boiler, tool or pipe?[17]

However, since a calculation of the Fair Value at the asset level is required for most valuation purposes, it is not sufficient enough to allocate the cash flow on production line level but an allocation on asset level is needed.

Apart from the difficulty of the identifiable cash flow the application of the income approach requires a risk adequate yield to discount the cash flow to the date of valuation. While this is regularly derived for an entire company using the Capital Asset Pricing Model (CAPM) method, it is extremely difficult to derive an appropriate discount rate for a single part of the company e.g. a production line. For an individual machine, system or tool, however, it is usually not possible to do this sensibly. This circumstance alone means that even if the second question can be answered positively, it is difficult in practice to derive a risk-adequate interest rate for discounting the separable cash flow of the asset.

In practice, therefore, not only the question of whether there are sufficient comparable transactions available, but also the question of the measurable cash flow at the level of individual assets will regularly have to be answered negatively. So, all that remains is a reference to the third approach, namely the cost approach.

Although the cost approach is the dominant method for the valuation of Machinery and Equipment, different methods will regularly be used within one valuation project. This is due on the one hand to the availability of data, and on the other hand to the different nature of the assets.

Land is usually valued with the market approach, standardized buildings often with the income- or cost approach and Machinery and Equipment with the cost approach. It is therefore entirely possible and usual to use all three methods for the different asset categories when valuing the tangible assets of a single production site.

While also different valuation approaches are possibly applied for one and the same assets, it is strongly advised not to carry out a valuation using two or all three of the approaches mentioned above in order to calculate an average of the results derived. Rather it is advisable to use the approach shown above to derive the best method.

The use of another method for plausibility purposes, however, is always welcome and advisable. If the primary approach chosen is the cost approach, it is advisable e.g. for Machinery & Equipment to check the value determined for at least certain individual assets, if available, using comparable purchase prices. In the case of Real Estate when choosing the cost approach as a primary approach to value a building, the determined asset value should at least be checked using comparable KPIs or the investment method. With a market rent derived, even if only approximately, the implied interest rate can be cross referenced with information from comparable markets or experience within a reasonable range.

This plausibility crosscheck is not always necessarily a comprehensive and complex valuation exercise. Sometimes a "back of the envelope" plausibility check with local markets KPIs or benchmarks can help to assess if the valuation result is reasonable and plausible.

6.6 Market Approach

The market approach, also known as the market data approach,[18] is a statistical approach by which the Fair Value is determined out of comparable market transactions[19] and, in certain circumstances, out of prices requested for Fixed Assets offered for sale. Provided that there are a sufficient number of comparable transactions available, this is a straightforward valuation method and is often considered the preferred method in the literature due to its direct link to the market.[20]

The market approach is based on the principle of substitution whereby a typical buyer will not purchase a property at a price higher than that of similar asset or properties with comparable locations, characteristics, and utility. The Appraisal Institute defines the sales comparison approach as:

"A set of procedures in which a value indication is derived by comparing the property being appraised to similar properties that have been sold recently, applying appropriate units of comparison, and making adjustments to the sale prices of the comparables based on the elements of comparison."[21]

Compared to other sophisticated methods, the basic scheme and the necessary calculations are relatively simple. For Machinery and Equipment, the price can be compared directly with the object. For Real Estate, the size of the property multiplied by the adjusted price per square meter gives the value of the property.

The market approach basically comprises of two different methods: the direct sales comparison approach and the indirect sales comparison approach (Figure 6.5). While the methodological approach is similar the application is basically driven by the market data available and potential adjustments to be made on those comparable data.

The sales comparison method is generally applied for assets, which are either traded in markets with features close to an active market or assets for which prices from recent transactions of comparable assets are available.[22]

6.6.1 Direct Sales Comparison Method

The direct sales comparison approach derives the value of a property directly from observed sales prices of comparable properties.[23] The direct comparative value method is used so frequently and, above all, as a matter of course in property analysis, that it is often forgotten that it is an independent method. The method can be used independently or be part of a more complex valuation method.[24]

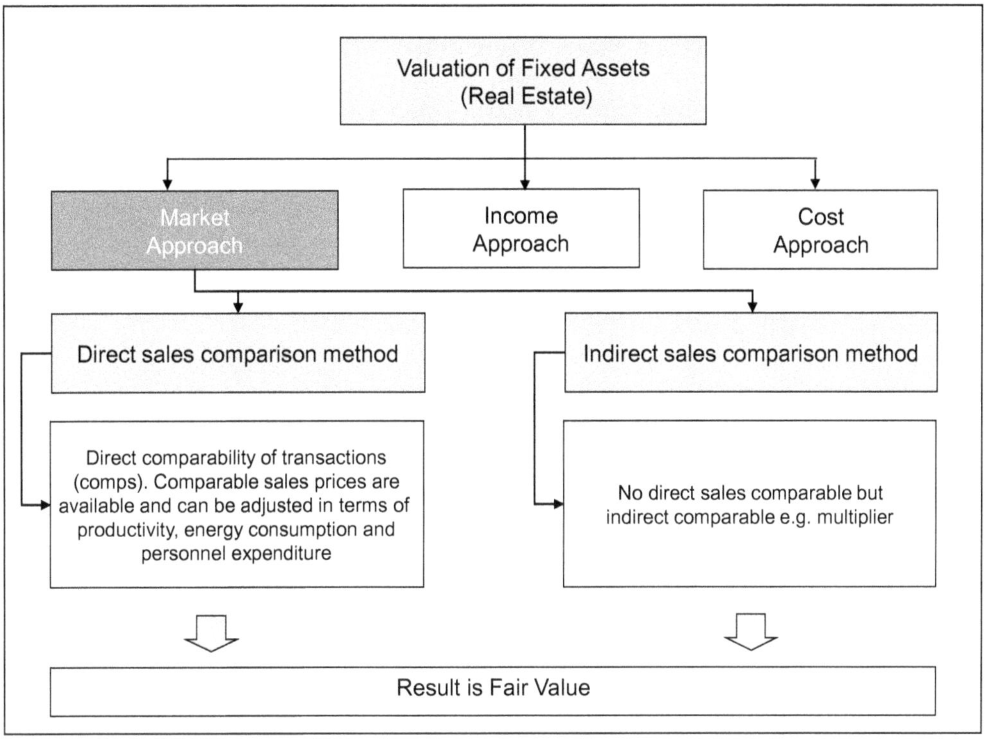

Figure 6.5 Valuation of Real Estate, Market Approach.
Source: Authors.

For the direct sales comparison method to be applied effectively, there must be a sufficient number of comparable sales prices available, the key characteristics of the comparable properties should closely align with those of the asset being valued, and the comparable transactions should occur near the valuation date. If these conditions are not met, the direct sales comparison approach cannot be applied. In such cases, deviations should be accounted for through adjustments in an indirect sales comparison approach.[25]

What a "sufficient number" means is often not clarified in literature or regulations. It should be more than one and generally the more the better. However, even more important than the amount of comparables is their quality in terms of comparability. Generally, it can be argued that a limited number of "good" comparable transactions which are similar in characteristics to the valuation object are to be preferred over a larger number which is merely comparable.

Because especially for Real Estate no two properties are truly identical, it is frequently impossible to find comparable sales. Even those sales judged roughly comparable ordinarily require adjustments for an indication of the subject's value. Therefore, the sales comparison approach will rely upon the analysis of recent transfers of similar properties, which in turn are compared and contrasted with the subject. While a wide range of factors may affect comparability, the intent is to establish a plausible range from which conclusions can be drawn and trends discerned.

6.6.2 Indirect Sales Comparison Method

In the indirect sales comparison approach, the value of a property is not derived directly from comparable transactions but indirectly from average comparable or comparable KPIs such as published average standard land values or earnings multiples.[26] These values must first be adjusted to the valuation date and/or to the specific characteristics of the property being assessed.

Key characteristics influencing value include the property's location, existing rights and encumbrances, the type and extent of permissible use, the shape and size of the plot, its development status, and soil quality. Additional value reductions may arise from burdens such as outdated structures, demolition orders, or other regulatory requirements like heritage protection.[27]

If the comparable prices are not derived from transactions conducted near the valuation date, they must be adjusted to account for temporal differences e.g. by using index series, to accurately reflect the valuation date.[28]

6.6.3 Valuation of Vacant Land

In practice the Market Approach is a widely utilized method for valuing vacant land, favored for its reliance on actual market transactions. This approach involves comparing, analyzing, and relating sales data of similar land parcels to the land being appraised. The highest emphasis is placed on sales of comparable properties that occurred close to the valuation date and under similar economic conditions (direct sales comparison).

When determining the value of land, purchase prices of comparable properties are crucial. Features influencing the land value must be considered, allowing for adjustments such as increases or reductions. These features include the property's location, rights and encumbrances, permissible development type and extent, shape, adjacent public services, and soil quality. Depreciation factors such as soil contamination, obsolete structures, demolition orders, or other legal constraints like preservation orders may also affect the land's value.

In certain territories local authorities also provide average prices or benchmark land values. Such benchmarks may also cater for a valuation of land (indirect sales comparison).

6.6.4 Market Approach for Machinery and Equipment

As mentioned already, for Machinery and Equipment assets, especially very specialized assets, the sales comparison method is seldomly applied due to the lack of sufficient market information as well as intensive research efforts needed when valuing large portfolios of assets.

As outlined in Figure 6.6 as for the property valuation also for the Machinery and Equipment valuation, the market approach also comprises two methodologies:

- The direct sales comparison method, and
- The indirect sales comparison method.

In practice the market approach for Machinery and Equipment assets is frequently applied to either valuing large individual assets or when valuing large quantities of assets to crosscheck the results derived from other methodology on a sample basis.

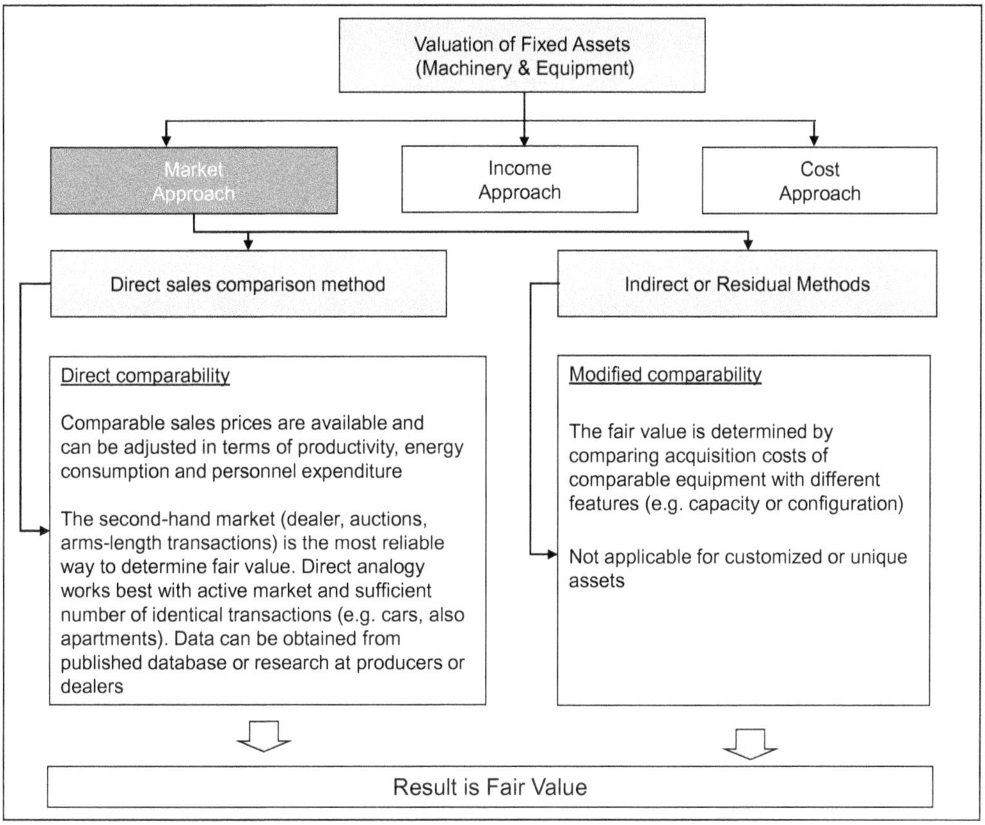

Figure 6.6 Valuation of Machinery and Equipment, Market Approach.
Source: Authors.

6.6.5 Direct Sales Comparison Method for Machinery and Equipment Assets

The direct sales comparable method establishes value based on a direct match of the subject to an identical asset or comparables.

This approach is applicable in practice for highly standardized Machinery and Equipment assets, where direct comparables may be observable in relevant markets.

Those standardized assets typically include commodity asset classes such as vehicles, forklifts, computer equipment or frequently traded standard production machinery assets. For cars there are regularly available online databases which show benchmarks for used cars and trucks and are based on actual transactions reflecting comparable.

Potential adjustments have to be made to reflect differences in age, capacity, efficiency and condition. In practice, those adjustments may contain a certain element of subjectivity. Therefore, the more quantitative those adjustments can be substantiated and the more transparent they are derived, the better.

In many cases such data for actual transactions of used assets might not be available and the valuer might use offer prices as only reference point to derive a comparable. Those offer prices might be found on special internet trading platforms or auction houses. When applying offer prices it is important to understand that this offer prices only reflect the sellers view on the price. An individual offer price might therefore not be considered a sufficient market comparable, but further adjustments might have to be applied. To bridge the gap between the asking price of the seller and a reasonable market price those adjustments are even more difficult to derive, since the element of subjectivity can be greater.

This is even more true for an inactive market and/or when assets are highly specialized. An indication of the reasonability of the price offered by the seller might be the period of time in which the asset is already offered on the market. If an asset is offered on the market, without any potential purchaser being attracted, this might be an indication that the price is too high, and a substantial downward adjustment needs to be considered. If on the other hand the asset is sold very fast on the market, this might indicate that the offer price was too low. For Real Estate, those analyses are performed and sometimes published by large online platforms, but those analyses might be rarely available for specialized Machinery and Equipment assets. Moreover, the marketing period is not only related to the asset price offered but the overall market sentiment and situation in the specific industry as well.

On the other hand, it has to be considered that special assets require specific buyers. Hence, the more specialized an asset is, the longer the sales process might take.

Considering the outlined above, offer prices might be a helpful reference point when researching sale comparable to derive the value of Fixed Assets but must always be considered with particular caution.

6.6.6 Indirect Sales Comparison Method for Machinery and Equipment Assets

The indirect market comparison method involves estimating the Fair Value of an asset by analyzing comparable assets that share similar, though not identical, characteristics. This approach utilizes various measures of utility, such as size and capacity, as the basis for comparison. For instance, when evaluating the value of an engine produced by Company A, the appraiser may not find any sales data for engines from Company A but may identify sales data for engines from Companies B and C. This type of comparison is inherently more subjective than a direct comparison, necessitating further adjustments through a detailed analysis of the comparability elements involved.[29] A good example is a vehicle valued using a published pricing guide.

Another aspect must be considered as well when valuing assets by comparing them to individual market transactions. In most cases the valuation of an asset is not to be seen on an individual (isolated) level but as part of an ongoing production process: Going Concern. This is in most cases less important for Real Estate assets but more important for M&E assets. While the single asset might be of less use if viewed individually or even might be seen as Scrap Value, it might have significant value to the production process contributing as part of the entire business. This needs to be taken into consideration when valuing assets by comparing them to market transactions as well.

Moreover, often individual, perhaps even very high-quality tools are only used rarely in production but are then essential for production and thus contribute to the overall success of production and the company. If these perhaps very expensive assets were assigned a lower value due to their rare use, this would certainly not be appropriate.

6.6.7 Summary and Practical Application of the Market Approach

To apply the sales comparison method, the steps shown in Figure 6.7 have to be performed.

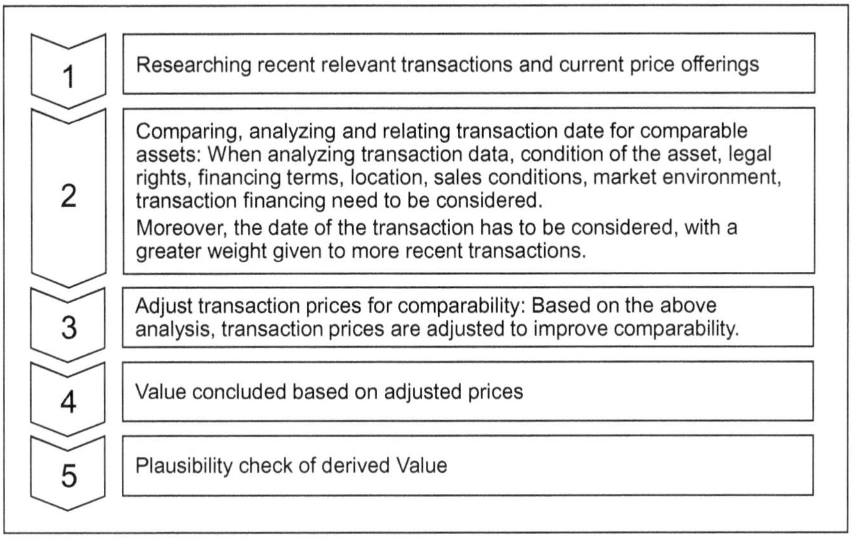

Figure 6.7 Practical Steps for the Application of the Market Approach.
Source: Adapted from ASA [2011], p. 93.
Notes: See also, Corge et al. [2001], p. 308.

6.7 Income Approach

The income approach is a valuation approach based on investment calculation.[30] The value of the asset is determined by the future surplus income that can be generated with it, which is discounted to the valuation date.[31] The value is therefore the result of the interaction between income and yield (Figure 6.8).

This ratio of income and yield is the simplest form of the income approach, also known as the investment method. While it cannot be used for complex valuation issues, it is nevertheless useful for quickly and easily determining approximate values or checking the plausibility of valuation results.[32]

The gross initial yield can be derived from the converted formula, and the net initial yield can then also be derived by taking outgoings and incidental transaction costs into account. The net initial yield is one of the most important key figures in property investment and is regularly published as a KPI (Figure 6.9).

The three parameters or Items: Income, Yield and Value represent the basis for the income approach. All other methods essentially rely on this fundamental principle.

6.7.1 Mathematical Basics for Income-Based Valuations

The financial mathematical principles of the income approach primarily refer to the most basic methods from investment calculation.

The time value of money plays a central role in all medium-to-long-term investment projects, with future payments being estimated to be lower than current ones. "A dollar today is worth more than a dollar tomorrow."[33]

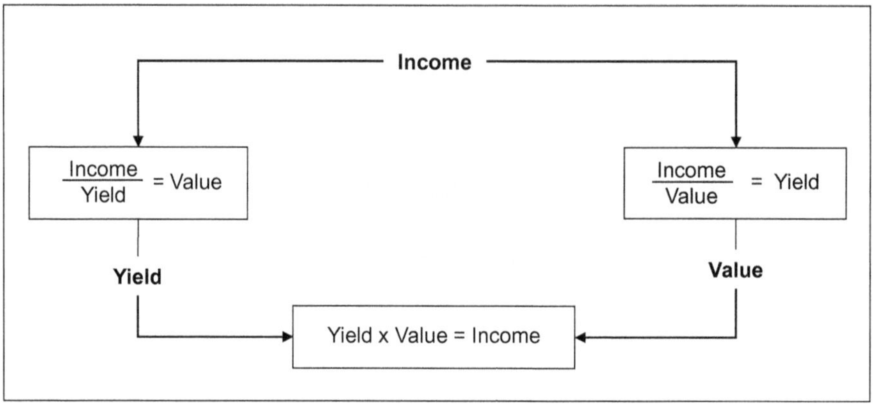

Figure 6.8 Basic Principles of the Income Approach.
Source: Authors.
Notes: Kleiber [2020], p. 1317; see also Corgel et al. [2001], p. 322.

Purchase Price	**150,000 USD**
Transaction Cost (approx. 12%) for agent, land transfer tax, land register, notary	18,000 USD
Investment Cost	**168,000 USD**
Gross Income p.a.	**7,500 USD**
Maintenance Cost	-600 USD
Administration Fees	-250 USD
Net Operating Income	**6,650 USD**

Gross Initial Yied (GIY)

$$GIY = \frac{\text{Gross Income p. a.}}{\text{Purchase Price}}$$

$$GIY = \frac{7{,}500}{150{,}000} = 5\%$$

Net Initial Yield (NIY)

$$NIY = \frac{\text{Gross Income } - \text{Outgoings}}{\text{Purchase Price} + \text{Transaction Costs}}$$

$$NIY = \frac{6{,}650}{168{,}000} = 4\%$$

Figure 6.9 Gross Initial Yield and Net Initial Yield.
Source: Authors.
Note: Cf. Kleiber [2020], p. 1317.

A distinction is made between the present values of three different payment flows or types of payment (see Figure 6.10). There are three fundamentally different scenarios for discounting or capitalizing future payments.

1 The "perpetual annuity": An open-ended cash flow (I) is capitalized with a risk-adequate interest rate (Y) and the value (V) is determined on the valuation date.

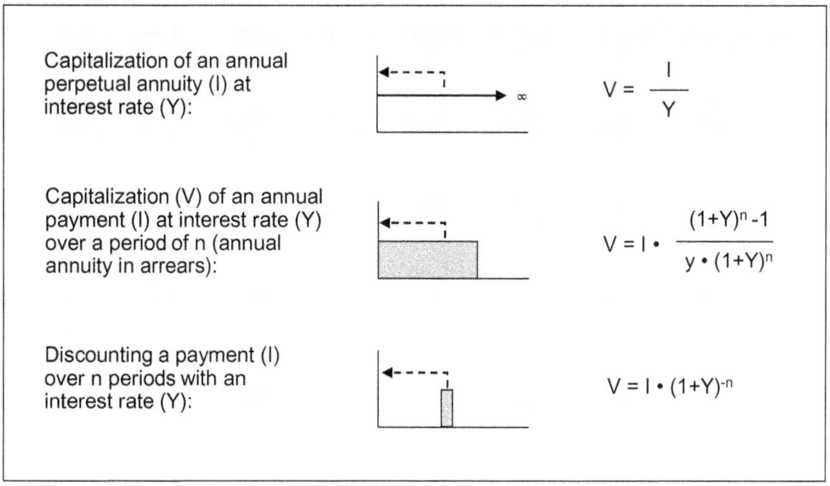

Figure 6.10 Financial Mathematic Basics.
Source: Authors.
Notes: See also White et al. [2003], p. 79; Brealey and Myers [2000], p. 41; Metzger [2018], p. 175.

2 In the second case, the annual cash flow is limited in time. The annual payment (I) is discounted over the limited period (n).
3 In the third case, it is not a multi-year cash flow, but a one-off payment (I) in the future (n) that is discounted to the valuation date.

All internationally common income-based valuation methods are based on either one of the three formulas or on a combination of the three formulas. Depending on whether it is a static method, the perpetual annuity will be used, or in the case of dynamic methods in which different payments are planned in the years, the second or third formula will be used.

In addition to income, which in the case of Real Estate investments is usually rent generated from the property, yield is the second input variable that determines valuation.

The level of yield depends on the risk associated with receiving the future cash flow. The higher the risk, the higher the yield and the lower the present value of the payment.

This applies not only to property, plant and equipment, but also to the valuation of all assets and companies.

As outlined earlier, in the course of fixed asset valuation projects the income approach is mainly applied for the valuation of Real Estate related assets. The income approach is primarily used for properties where the focus is on "income orientation" or for properties for which there are rental agreements or comparable rents can be obtained from the market. It is therefore used in particular for rental properties, business premises and mixed-use properties, but can also be used for production buildings or warehouses. Therefore, the major focus of the income related valuation chapter is related to the property valuation.

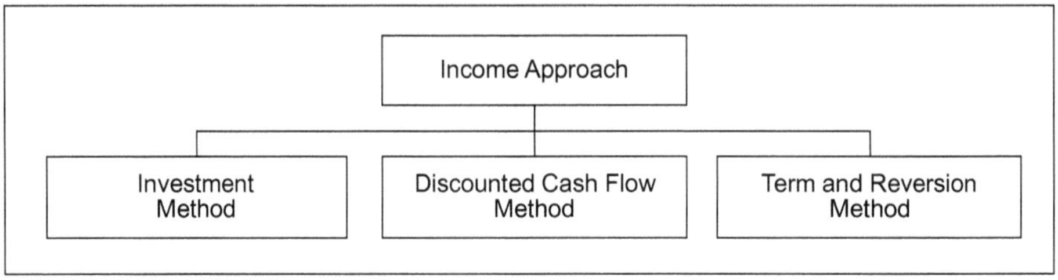

Figure 6.11 Income Approach.
Source: Authors.
Notes: Cf. Wyatt [2013], p. 135; Manganelli [2015], p. 191.

6.7.2 Income-Based Valuation Methodologies

The income approach comprises several different valuation methods, which largely vary in their complexity and are applied in different and not uniform ways.[34] However, as shown in Figure 6.11, in international valuation practice the most commonly applied income-based valuation methods include: the Investment Method, (also known as Direct Capitalization Method), the Discounted Cash Flow (DCF) Method and the Term and Reversion Method.

The three valuation methods differ considerably in their methodological complexity and level of detail. The static Investment Method is a very simple valuation method with only limited major assumptions. The dynamic DCF Method, on the other hand, is much more complex and can therefore also reflect much more complex valuation assumptions. The Term and Reversion method is somewhere between the two in terms of its complexity.

The method, which is to be used for the valuation in each individual case is decided not only by the reason for the valuation and the property, but also by the existing rental agreement situation.

As property markets move cyclically, the market rent for a property changes over time. As the rental agreement date is sometimes a long time in the past and not the valuation date, the actual rent agreed in the contract and the market rent that can be achieved on the valuation date can differ significantly. This leads to three different scenarios in property valuation (Figure 6.12).

Due to the inherent volatility of the Real Estate markets, the rack-rented property case described will in reality be the exception rather than the rule, and the rental income generated from the property will rarely correspond to the market situation on the reporting date.[35] In the case of so-called "reversionary properties," a distinction must be made between different rental levels when evaluating them. Firstly, the paid rental level (actual rent) and secondly, the future rental level (market rent) after the next rent adjustment (rent review). This leads to the need for so-called "two income models," whereby two different starting situations can basically arise.[36]

Generally, the income approach can be applied for all Fixed Assets. However, as described above, since income streams are difficult to separate and directly allocated to Machinery and Equipment assets, the income approach is less applied to Machinery and Equipment but mainly to Real Estate. Therefore, in the following description of the income approach the focus is more on the valuation of Real Estate related assets.

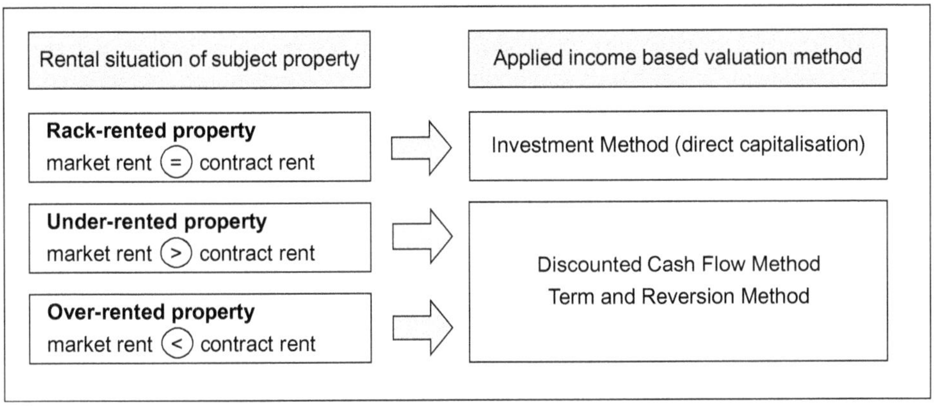

Figure 6.12 Rental Situation of Properties.
Source: Authors.

If the property is rented out below the current market level on the valuation date, this is referred to as an "under-rented property" case. This circumstance can be due to increased market rents since the contract was signed, inflationary developments or one-off payments made to the landlord at the beginning of the contract term.[37]

If the contractually agreed rent is above the market rent, this is an "over-rented property" situation, where it can be assumed that the current rental income cannot be generated on the market beyond the contract term. In practice, this case often occurs when there are negative economic developments and the tenant's forecasts are too positive when the contract is signed.[38]

While the static investment method can be used for rack-rented property, dynamic methods such as the DCF method are used.

Generally, the more complex a property is e.g. several different usages, lease terms, tenant incentives, the more complex the applied valuation method and related valuation model has to be in order to reflect such complexity in the valuation. Nevertheless, it should not be forgotten that more parameters increase complexity and all future assumptions are only forecasts. In this context, Damodaran points out that complex methods are not always necessarily better than simple methods because each additional input parameter is often based on assumptions made by the valuer.[39]

It is therefore necessary to decide on a case-by-case basis which level of accuracy is reasonable and which scenarios depicted may be pseudo-accuracies. Therefore it's advisable to choose a method and related model "as simple as possible but as complex as necessary."

6.7.3 Derivation of Parameters

Compared to the Investment method as well as the Term and Reversion Method, the DCF Method allows for a much more detailed scenario planning of the future Income Streams / Cash Flows. This leads to more parameters and related assumptions.[40]

6.7.3.1 Income Stream (Rental Income)

The income stream consists of payments from renting out space and other sources (e.g. billboards, mobile phone antenna, solar panels). The forecast for future rental payments is primarily based on existing lease agreements such as fixed rent, graduated rent, turnover rent or index-linked rent. If the lease does not specify such terms, statistical data or empirical values can support a standardized forecast (indexation), considering the regional supply and demand situation and legal regulations.[41]

For vacant rental spaces and those with expiring leases, assumptions must be made regarding future rentability. In such cases, the rental income should reflect the market rate achievable at the time of new or renewed leasing.

6.7.3.2 Non-Recoverable Costs (Outgoings)

The Non-Recoverable Costs are also frequently referred to as outgoing expenses or outgoings that need to be forecasted, including administrative costs, maintenance costs, and non-recoupable operating costs. This forecast can be made by updating actual values or using industry-specific empirical values. These values can be expressed in absolute terms or as percentages (e.g. maintenance costs as a percentage of rental income). Maintenance costs should be derived from expected values, maintenance plans, or calculated amounts, taking into account factors such as the building's use, year of construction, construction type, and quality. It's also important to consider potential rent reductions and temporary vacancies due to tenant turnover and renovations.

6.7.3.3 All Risk Yield

The All Risk Yield (ARY) reflects the risk associated with the projected future income of the asset[42] and can be basically derived by two different ways. Either the so-called build up method or derived implicitly from comparable market transactions.[43]

The build-up method (Figure 6.13) is based on a risk-free rate, e.g. 10-year government bond. On this rate which already reflects the country risk profile, additional adjustments for the specific risk for the asset class e.g. Real Estate and the specific asset need to be added.[44]

The risk for the asset class reflects the systematic risk inherent in the asset class. The unsystematic risk reflects the individual risk associated with an investment in the subject property.[45] While the risk-free rate is in practice easy to derive, it is more difficult for the industry risk and even the asset risk. There might be indications available for risk spreads from financial capital markets but to derive the adjustment factor for the individual property is difficult. Hence in practice the All Risk Yield is typically derived from transactions in the market and observable market rents using the Net Initial Yield (NYI) formula. It is also frequently published in local market reports.

6.7.3.4 Discount Rate and Capitalization Rate

In the DCF Method a discount rate is applied to derive the Net Present Value of payments from the holding period while the Terminal Value is calculated by using a Capitalization rate. The derived Terminal Value then is discounted to the valuation date by applying the discount rate.

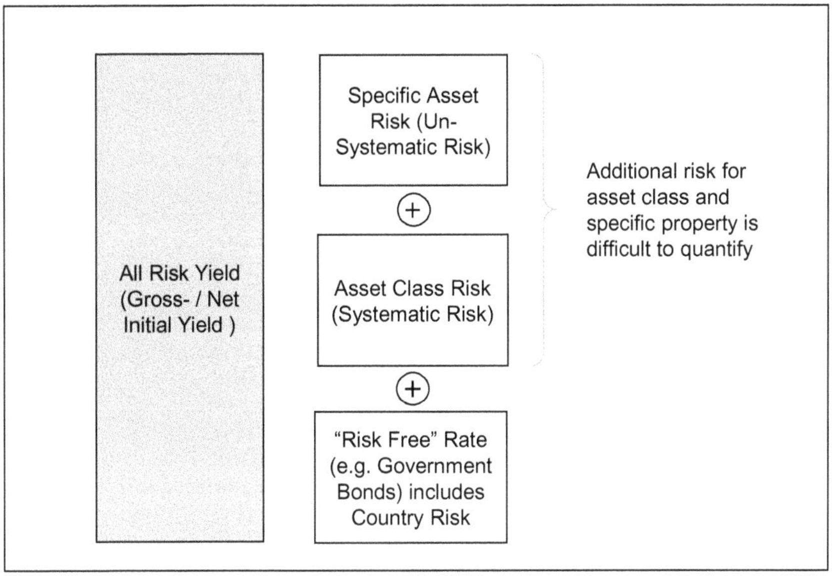

Figure 6.13 Build-up Method.
Source: Adapted from White et al. [2003], p. 31; and Kleiber [2018], p. 1443.

The discount rate is the measure of the average market interest rate on the capital that is usually tied up in a property for a long period of time like the ARY. It is made up of a (risk-free) base interest rate and risk premiums and is also a measure of expected future increases in value. In valuation practice, however, due to the difficulties described above, it is mostly determined deductively based on market analyses rather than cumulatively.

The discount rate basically depends on the type of use, location characteristics, rent and remaining useful life (RUL), among other things, and together with the capitalization interest rate applied, it has a significant influence on the property value.

Growth and inflation are usually taken into account directly in the cash surpluses within the detailed planning period. Therefore, the discount rate which is applied to the annual income surplus for each year in the holding period is a nominal yield not including growth deductions. For the Terminal Value, which is derived by the static Investment Method, it is the other way round, growth cannot be considered in the income stream but only in the capitalization rate through the so-called growth discount (Figure 6.14).

6.7.4 Investment Method

When applying the investment method, the Market Value of a property is determined by capitalizing the forecasted net cash flow's Net Operating Income (NOI) using a rate of return (perpetual annuity) that reflects all risks associated with the property (All Risk Yield).

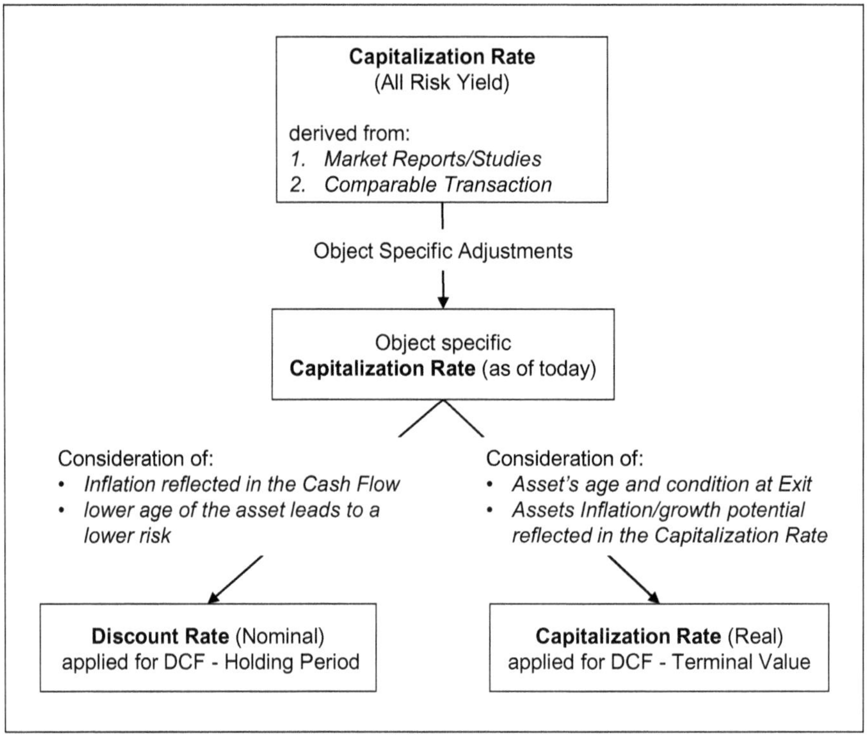

Figure 6.14 Derivation of Discount Rate and Capitalization Rate.
Source: Authors.
Note: Also see: Kleiber [2018], p. 1444.

Investment method

$$V = \frac{NOI}{ARY}$$

V Value (Market Value/Fair Value)
NOI Net Operating Income
ARY All Risk Yield

Figure 6.15 Mathematical Equation for Investment Method.
Source: Adapted from White et al. [2003], p. 31.
Notes: See also Brealey and Myers [2000], p. 17.

This method is applied by converting a single year's net market rental value expectancy into an indication of value through the direct capitalization (Figure 6.15).[46]

The net market rental value is defined as the annual rent which can be realized in the market as at valuation date (Figure 6.16). The application of the Investment Method requires either a rack-rented subject property or a vacant property which can be rented out at market conditions.

The net initial yield is the net market rental value from a property investment on the date of acquisition annualized and expressed as a percentage of the property's value.[47]

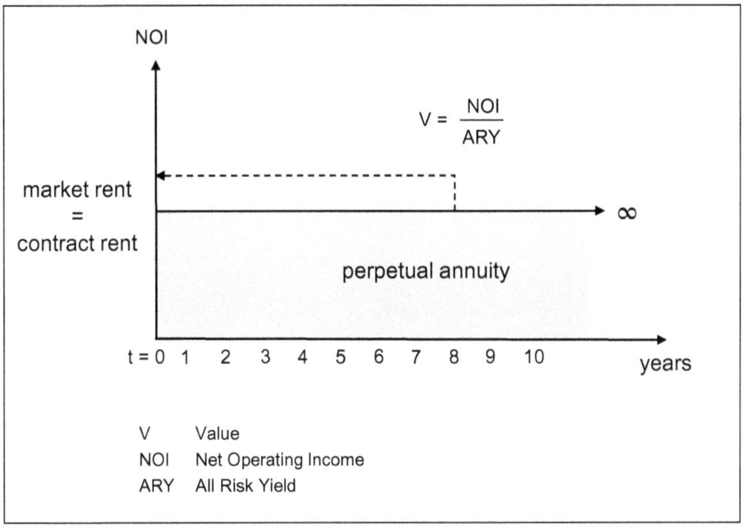

Figure 6.16 Investment Method.
Source: Authors.

The value can then be adjusted for the possible various nuances (adjustment factors) of the subject property which may vary from the "as is" value of the property.
The investment method should not be used if the property to be valued is not vacant or fully let at market rent at the time of valuation and there are tenancy agreements in place that are above or below market rent. Instead, the term and reversion method can be used to take into account the different tenancy agreement situation up to the end of the tenancy.

6.7.5 Term and Reversion

In contrast to the simple investment method, the Term and Reversion method distinguishes between the current rental period and the subsequent rental period. As shown in Figure 6.17, this allows contractual rents that deviate from the observable market rent as of valuation date to be represented precisely and rental contract situations to be taken into account in the valuation.[48]

If the current contractually agreed rental income deviates from the market-standard rental income, the surplus can be differentiated between a temporally limited part (term) specified by the lease agreement and a subsequent indefinite part (reversion) of the cash flow.[49] The capitalization of the first part is based on the remaining lease term applicable at the valuation date, while the present value of the remaining part is to be discounted over this remaining term.[50]

For under-rented properties Interest rates for the term are generally lower than the All-Risks Yield, because rental payments during the term are contractually secured, providing a high certainty of cash flow realization since the tenant is unlikely to terminate the other before the end of the lease term, and there is an expectation of rising rents in the future. Conversely, for over-rented properties, the interest rate for the term might be higher than the ARY, due to the potentially higher risk of realizing the cash flow and concerns about tenant creditworthiness.

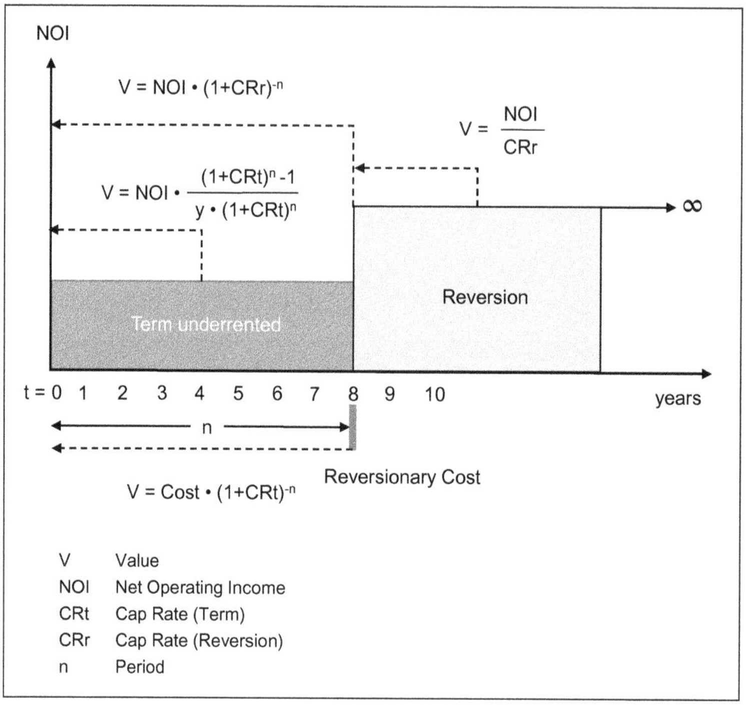

Figure 6.17 Term and Reversion.
Source: Authors.
Note: According to White et al. [2003], p. 99 and Leopoldsberger et al. [2016], p. 464.

In addition to the Outgoings so-called cost to reversion or reversionary cost may need to be considered. These costs are allowances for the related rent out after the termination of the current lease contract and the potential renovation or fit out for the new tenant.

The cost needs to be calculated individually depending on the type and usage of the asset as well as condition of the asset and it has to be discounted to the valuation date and deducted from the Net Present Value of the income.

Figure 6.18 shows an illustrative sample calculation applying the term and reversion method.

The Term and Reversion Method is typically applied for properties with one or a very limited number of existing lease contracts which differ from the observable market rent as of valuation date. If there is a larger number of contracts in place or the expected cash flow is to be volatile, the DCF Method is a better choice.

6.7.6 Discounted Cash-Flow Method

The DCF method is considered the most complex valuation method, allowing for complex valuation scenarios to be considered. It is a dynamic two-stage financial model used to calculate the present value of a property's future income, which is considered

Sample Calculation Term and Reversion

Term		**Reversion**	
Gross rental income	1,500	Market rent	1,200
Non recoverable cost	500	Non recoverable cost	500
NOI	1,000	NOI	700
Lease term	11 Y.	Capitalization rate (Reversion)	5%
Capitalisation rate (Term)	6%	Cost to reversion	(360)

Term

$$T = (GOI - NonRec) \bullet \frac{(1+i)^n - 1}{i \bullet (1+i)^n} \quad T = (1500 - 500) \bullet \frac{(1+6.00\%)^{11} - 1}{6.00\% \bullet (1+6.00\%)^{11}} = 7{,}890$$

Reversion

$$R_{t=n} = \frac{(GOI - NonRec)}{i} \quad R_{t=n} = \frac{(1{,}200 - 500)}{5{,}00\%} = 14{,}000$$

$$R_{t=0} = R \bullet (1+i)^n \quad R_{t=0} = 14{,}000 \bullet (1+5.00\%)^{11} = 8{,}330$$

Term & Reversion

$$T\&R = T + R_{t=0} - RC \quad T\&R = 7{,}890 + 8{,}330 - 360 = 15{,}860$$

Figure 6.18 Term and Reversion Sample Calculation.
Source: Authors.

its Market Value.[51] This involves creating a detailed forecast of revenue and expenses over a so-called "detailed planning period" or "holding period" of typically 10 years.[52] The holding period must generally be chosen so that the net income at the end can be considered stabilized i.e. unaffected by special circumstances that affect income. The forecast is based on current contractual agreements and anticipated market trends. At the end of the holding period, the net income is capitalized in perpetuity to estimate a hypothetical purchase price (Terminal Value).[53] The annual income and the purchase price are then discounted back to the valuation date to determine the property's Market Value.[54]

As shown in Figure 6.19 as well as the related mathematical equation Figure 6.20 the sum of the Net Present Values of each year's NOI of the Holding Period and the Net Present Value of the Terminal Value leads to the Value of the property or asset.

In both methods, neither the land value nor the RUL of the assets is explicitly taken into account; both influencing factors are included in the respective interest rates (all risk yield or discount and capitalization interest rate). These interest rates also usually reflect the net return.[55] In these cases, the result of the present value calculation must be reduced by the typical acquisition costs.

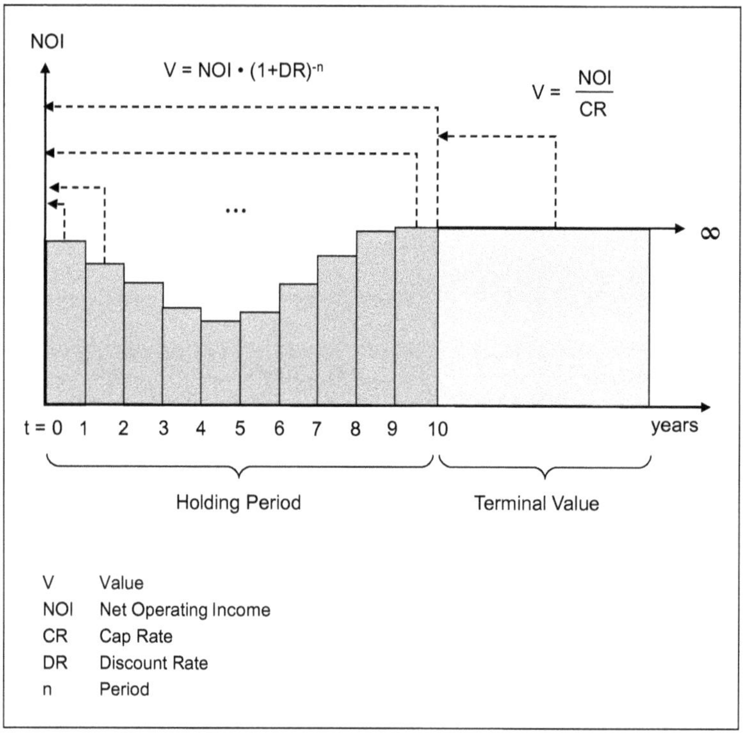

Figure 6.19 DCF Method.
Source: Authors.

Discounted Cash Flow Method

$$MV = \sum_{t=1}^{n} (GOI - NonRec)_t \times (1+DR)^{-t} + \left(\frac{(GOI\text{-}NonRec)^{t+1}}{CR} \right) \times (1+DR)^{-t}$$

V	Value (Market Value/Fair Value)
DR	Discount Rate
CR	Cap Rate
GOI	Gross Operating Income
t	Period
NonRec	Non Recoverable Cost / Outgoings

Figure 6.20 Mathematical Equation for DCF Method.
Source: Authors.

This method is only suitable for deriving the Market Value if market-derived, objectified costs are used. The increased uncertainty of the method due to the large number of assumptions to be made and the nature of the underlying calculation process (difference method) must also be taken into account.[56]

The advantage of the DCF method is that it is much more transparent and assumptions can be chosen on a much more detailed level than other traditional methods such as investment method or term and reversion method.[57]

Of all the income approach methods described, the Terminal Value approach is used in financial mathematics to model the perpetual use of a property. For properties with very long remaining economic lives, this assumption has no significant effect due to the low present value of future payments. The situation is different for properties with a short RUL. Here, the actual RUL must be considered by adjusting the interest rate, and the influence of land value on property value increases. This must be taken into account in all income approach methods when valuing properties with short RULs.

At the end of the holding period, the Terminal Value of the property must be estimated if the economic useful life of the property extends beyond the detailed forecast period.[58]

The Terminal Value is generally determined by capitalizing the excess cash flow from the last detailed period by implicitly applying the Investment Method. The derived Terminal Value must then be discounted to the valuation date.

When deriving the planning period used as a basis for the Terminal Value determination, it must be ensured that the corresponding excess cash flow is stable and therefore does not contain any unscheduled payments or withdrawals. The capitalization interest rate is determined depending on the Terminal Value model selected (infinite or finite annuity).

For the purpose of checking the plausibility of the valuation result, it is advisable to compare the initial yield of the valuation object (based on the first stabilized excess cash flow i.e. adjusted for extraordinary effects) with the initial yields usual on the transaction market for comparable objects.

6.7.7 Residual Value Method

The Residual Value method is a method primarily used by investors and property developers to determine the land price that is just about affordable for a development project.[59] This can involve undeveloped land or land with buildings scheduled for demolition or partly finished project developments.[60] Since the predicted value for the finished project is determined using the market or income approach, but the costs to be deducted are similar to the cost approach, the Residual Value method is a combination of several valuation methods (mixed method).[61]

Using the best possible usage concept that can be implemented within the framework of legal and economic possibilities, a hypothetical value of the finished property is first determined.[62] This can be done either based on anticipated comparative purchase prices or by capitalizing future rental income surpluses for the time after the property is finished. In a second step, all costs incurred in creating the property as well as a calculated development profit are deducted from this calculated developer value. The remaining difference between the fictitious sales price and construction costs is called the "residue" and corresponds to the value of the undeveloped land.[63]

The criticism of the residual method is primarily aimed at the deductive approach, the uncertainty of numerous input factors that cannot be determined precisely and the sensitivity of the results to small changes in these factors.[64] This mainly concerns the consideration of current rents and sales prices, although the value relationships may have changed

Assumptions

Holding Period:	10 years
Discount rate:	8.00%
Capitalization rate (=Exit cap rate):	6.00%
Rental Gross rental income:	100, growth rate: 1.50%
Non-recoverable costs (Outgoings)	10, growth rate: 2.00%
Purchasers' cost:	5.75%

Year	**1**	**2**	**3**	**4**	**5**	**6**	**7**	**8**	**9**	**10**	**11**
Gross rental income	100.0	101.5	103.0	104.6	106.1	107.7	109.3	111.0	112.6	114.3	116.0
Non-recoverable costs	-10.0	-10.2	-10.4	-10.6	-10.8	-11.0	-11.3	-11.5	-11.7	-12.0	-12.2
Net operating income	**90.0**	**91.3**	**92.6**	**94.0**	**95.3**	**96.7**	**98.1**	**99.5**	**100.9**	**102.4**	**103.8**
Exit value (Cap rate @6%)										1,730.0	
Discount factor @8%	0.93	0.86	0.79	0.74	0.68	0.63	0.58	0.54	0.50	0.46	
Discounted annual CF	83.7	78.5	73.2	69.6	64.8	60.9	56.9	53.7	50.5	47.1	
Discounted cashflow (sum)	**1,434.7**									795.8	
Purchasers' costs	-78.0										
Net value	1,356.7										
Say	**1,350**										

Figure 6.21 Sample Calculation of the DCF Method.
Source: Authors.

significantly by the time the project is completed. On the other hand, the exact estimation of all costs incurred in realizing the project is often difficult in practice. As a result, this can lead to either a land price that is too high or the forecast developer profit not being realized.[65]

The following simplified example illustrates the sensitivity of the residual approach: If the investment costs for a Real Estate project are calculated at USD 90m and the investment value (potential sell price) at USD 100m, the profit is USD 10m. However, if these assumptions change by just 5%, which is negligible in terms of construction costs, quantity surveying and valuation, the development costs would be USD 95m and the investment value would be USD 95m as well, resulting in a profit of zero. Therefore, extreme caution must be exercised when applying the residual method, and sensitivity analyses should be performed.

Despite its obvious risks in determining the initial parameters, the method is popular with property developers and investors due to its methodological simplicity or when there are insufficient comparable transactions.

6.7.8 Income Approach for Machinery and Equipment

As outlined above, the income approach is not widely used for the valuation of Machinery and Equipment; the reasons given include the difficulty in determining income that can be directly and separately related to a specific asset, the concern over the reliability of income forecasts, and the difficulty in deriving a risk-specific discount rate for the individual asset.[66]

This is particularly true when the asset or group of assets does not generate any income, such as when the asset is not currently in use, or when it's not possible to identify an income stream directly related to the assets. Often, the only identifiable income stream is associated with the business or on the level of production site or production line, rather than with a specific asset.

When it comes to the income approach in Machinery and Equipment in practice in most cases, only the DCF method is applicable for the valuation of Machinery and Equipment assets. The investment method assumes a perpetual income stream which can only be applied for assets with a very long useful life (like land and/or buildings) but rather not for Machinery and Equipment. The term and reversion is designed to consider rental agreement for a certain remaining time. Since Machinery and Equipment assets are seldomly rented out and less likely reflect an over- or underrated situation. Thus, the Investment method as well as the Term and Reversion method are not applicable in practice to Machinery and Equipment assets.

When applying the DCF Method for the valuation of Machinery and Equipment assets where an income stream is identifiable and can be discounted with an adequate discount rate, the valuation process is as shown in Figure 6.22.

The starting point for the Cash Flow is the income stream which is generated by the valuation subject (asset or group of assets). From this income direct and indirect expenses generated by the asset need to be deducted in order to derive the NOI. The NOI needs to be discounted to the valuation date with the discount rate.

The Net Present Value thus obtained (from the detailed planning period) is added to a possible Terminal Value and the Fair Value is determined.

For individual, very long-lived assets, a Terminal Value can be determined on a case-by-case basis. In most cases, however, due to their age, this will only represent the Scrap Value after the end of the detailed planning period.

If the DCF method is used to value an entire plant or production line that is regularly serviced and maintained, the useful life can be extended and, under certain circumstances, a Terminal Value can be derived based on an assumed perpetual use.

In practice, the DCF method can be used to value entire production facilities and large infrastructure assets, such as power plants, large refineries and petrochemical plants. If there are rental agreements for the infrastructure that can be verified as being in line with the market, these are used as the basis for underlying the cash flow. Otherwise, the company's cash flow can be used as an alternative basis. A certain proportion of turnover can then be recognized as a "notional rent" for the property, plant and equipment. While this can be a sensible approach in individual cases, it will often be difficult in practice to

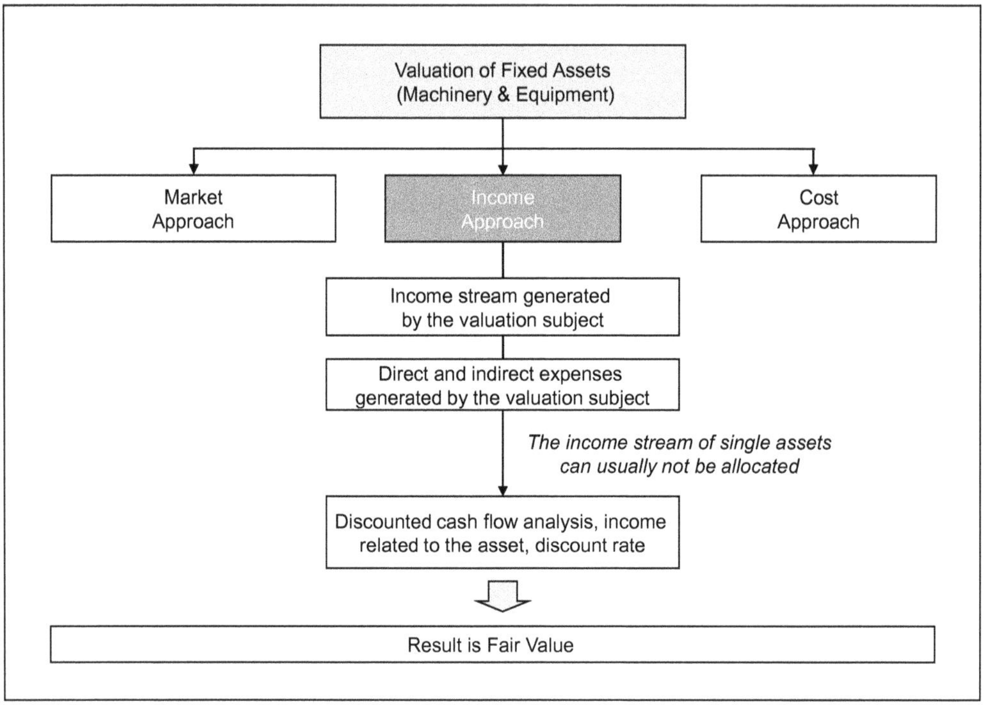

Figure 6.22 Valuation of Machinery and Equipment.
Source: Authors.

derive this rent as a percentage of turnover and justify its amount. Therefore, a transparent description of how the income stream was derived must be included in the valuation report.

In addition to the derived income another challenge in the DCF valuation of plant and machinery assets is, as described above, the derivation of a risk related yield to be applied. Unlike as for Real Estate, such key figures are hardly published for M&E assets and there is little empirical data available. An approach using the build-up method based on a risk-free base interest rate plus systematic and unsystematic risk premiums is theoretically conceivable, but in practice it is very difficult to quantify. One approach can be the approximation using the company-specific weighted average cost of capital (WACC) as a reference point. Although this provides information on the industry-specific risk, it reflects the company's risk, which does not necessarily correspond to that of the individual tangible assets. Caution is therefore advised here and the WACC approach for the valuation of M&E assets may be plausible in terms of the amount but is not correct in terms of valuation theory.

6.7.9 Summary and Practical Application of the Income Approach

There are nine basic steps in the income approach as shown in Figure 6.23.

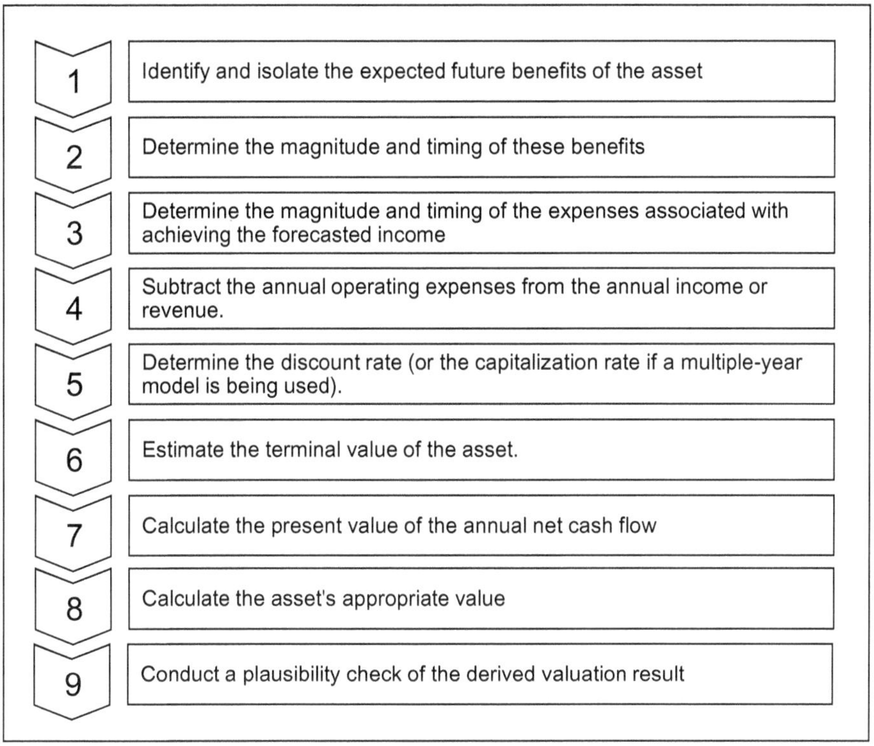

Figure 6.23 Practical Application of the Income Approach.
Source: Authors.
Note: Cf. ASA [2011], p. 132.

6.8 Cost Approach for Real Estate

The Cost Approach is a substance-oriented method in which the focus is less on future income than on the costs associated with the construction of the asset or property.[67]

The basic idea is that a potential buyer is not willing to pay more for a property than the opportunity costs of building it themselves.[68] The method is primarily used to value so-called "specialized properties" (e.g. schools, churches, hospitals) or industrial properties (e.g. oil refineries, chemical factories) for which there is no sufficient market activity due to their special characteristics.

The application of the cost approach is a two-part process, whereby the land value and the value of the buildings are first determined separately and then added to the total value of the property (Figure 6.24).[69] The value of the land is normally derived using the comparative value method or using the residual approach.[70]

The "depreciated replacement cost" used to determine the value of a building is defined in the Red Book as the costs that must be spent on rebuilding a comparable

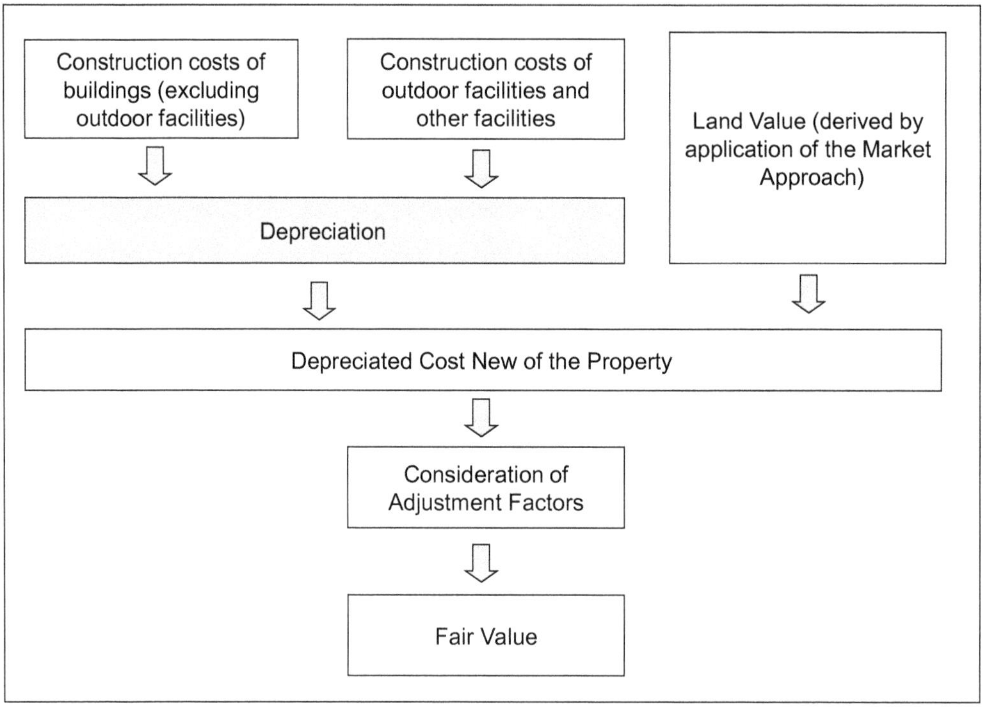

Figure 6.24 Cost Approach.
Source: Authors.
Note: Cf. Metzger Bernhard [2018], p. 120.

building. The construction costs are estimated on the basis of the latest appropriate construction techniques and legislation and specifications corresponding to the type of use. It is not necessary to rebuild the building as an exact copy with the same architecture, but rather to ensure equivalent performance and facilitate same usage.[71]

So-called "unit costs" can be used as the basis for calculating construction costs. These are usually area- or room-related (costs per square meter or cubic meter) but can also be used as comparative costs for other units e.g. per bed in hospitals or hotels. The application will be determined by the type of use and nature of the building. Although, if cost information per cubic meter is suitable for buildings with particularly high rooms such as warehouses or production factories, the most common case in practice will be the calculation using area-related cost rates.[72] The costs determined in each case must be corrected for age depreciation, taking special influencing factors into account, and the net replacement cost can be derived in this way.

The depreciation is determined by the RUL in relation to the total useful life of the building ("TUL"), whereby the RUL can be extended depending on the type and extent of repair and modernization measures (Figure 6.25). Conversely, if investment measures are not carried out (so-called repair or maintenance backlog), the RUL can be significantly shortened and be less than the TUL minus the age.[73]

Remaining Useful Life (RUL) = Total Useful Life (TUL) – Age

$$\text{Depreciation}^* = \frac{\text{Total Useful Life} - \text{Remaining Useful Life}}{\text{Total Useful Life}} \times 100$$

*Depreciation in % needs to be deducted from Cost New in order to derive its respective time value

Figure 6.25 Calculation of Depreciation.
Source: Authors.
Note: Cf. Metzger [2018], p. 176.

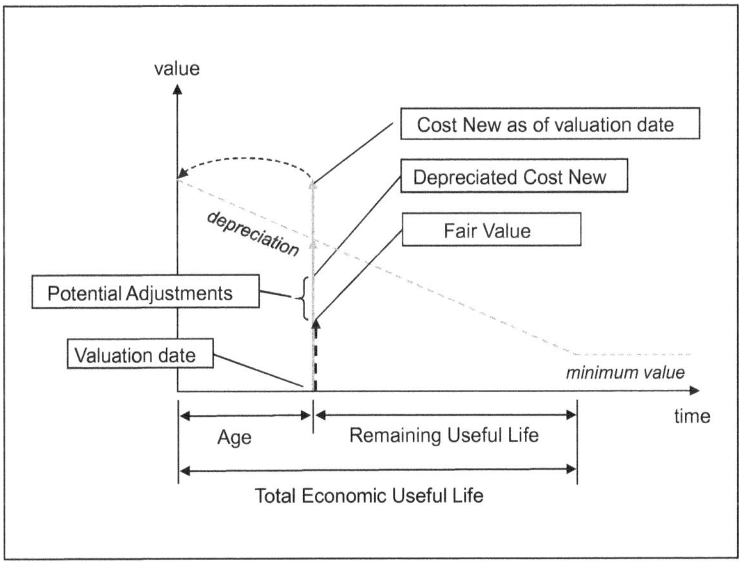

Figure 6.26 Depreciated Cost New Method.
Source: Plock et al. [2025], p. 13.

The RUL is considered as the period, for which, beginning with the record date of the valuation, an economic utilization of the building is to be assumed. As a rule it is shorter than the (technical) remaining life, which mainly depends on the construction of the foundation and the other supporting constructional structures.

The RUL can be calculated as the difference between the normal TUL of a building and its age as of valuation date (Figure 6.26). If necessary, this calculation is to be adjusted according to the state of repair of the building, the general and local economic conditions and the future exploitability (utilization demand) whereby in value determination practice an estimate of the RUL on the basis of these value criteria without an explicit consideration of the year of construction is increasingly accepted.

If the TUL due to proper maintenance of the building is still in good condition to facilitate its intended use, a so called "fictional construction date" can be applied. Hereby it is assumed that due to substantial reconstruction and modernization measures to the building its state of construction corresponds to a building constructed at a later date. In such cases the RUL and the depreciation due to age are not to be determined according to the actual age of the building but in accordance with the extent of the constructional renewal. As the TUL remains unaffected by this, this method results in a fictional year of construction.

The deductions for defects and/or disrepair correspond as a rule to the costs of maintenance and other construction work (e.g. reconstruction) necessary to bring about a state of repair to be expected of a building of this age to guarantee its utilization for the normal RUL period.

The method should only be used if no other methods can be applied as a primary valuation method, and the result should be crosschecked by benchmarks.

6.9 Cost Approach for Machinery and Equipment

When there is no sufficient market data available (second hand prices or replacement prices of comparable assets) to apply the market approach or the income approach, because of the specialized nature of the asset the asset should be valued using the cost approach.[74]

As for Real Estate, also for Machinery and Equipment the cost approach is based on the economic premise that a buyer is not prepared to pay more for an asset than the cost of acquiring an asset of equal value as of the valuation date. When applying the cost approach, it must be taken into account whether unreasonable time periods for a purchase or construction, barriers to entry or other regulatory or market obstacles, unreasonable risks or other factors may conflict with the application of the asset value method.

As mentioned above in Chapter 3, since Machinery and Equipment assets are often highly specialized, in practice, the cost approach will be the preferred approach in most of the projects.

6.9.1 Depreciated Cost New

When applying the cost approach, also frequently referred to as Depreciated Cost New method, the starting point is the current replacement cost, reproduction cost or cost of equivalent new of the asset (Figure 6.27).[75]

The RCN[76] is generally the starting point for developing an opinion of value using the cost approach. The RCN describes the cost on the valuation date at which a new asset can be installed in a new and in perfect condition in a contemporary design in terms of its function and parameters. The RCN includes not only the value of a machine e.g. but also all necessary ancillary costs, e.g. for freight, assembly, installation, foundation.[77]

From this cost new, loss in value due to physical deterioration and functional, economical as well as potential technical obsolescence needs to be deducted to derive the asset's Market Value.[78]

Even though the two terms Replacement Cost New and Reproduction Cost New are often used synonymously, they must be clearly distinguished from each other in terms of definition.

Replacement Cost New refers to the present expense of acquiring a new asset that offers the closest equivalent utility to the asset being appraised, emphasizing the replication of the asset's utility rather than its precise physical characteristics.

Reproduction Cost New denotes the current expenditure required to produce a new duplicate of the asset being appraised, utilizing the same or very similar materials. This process involves creating an exact replica of the original asset.[79]

Cost of Equivalent New describes cost to purchase the new available version of asset to replace the asset to be valued.

The RCN can be determined using the new price of Fixed Assets of the same type and performance as of valuation date. However, when using current comparative prices for assets, advances in production technology must be taken into account.[80] For successor systems of the same type, purpose and performance, discounts for technical innovations must be made due to technical progress, depending on the ratio of the technical

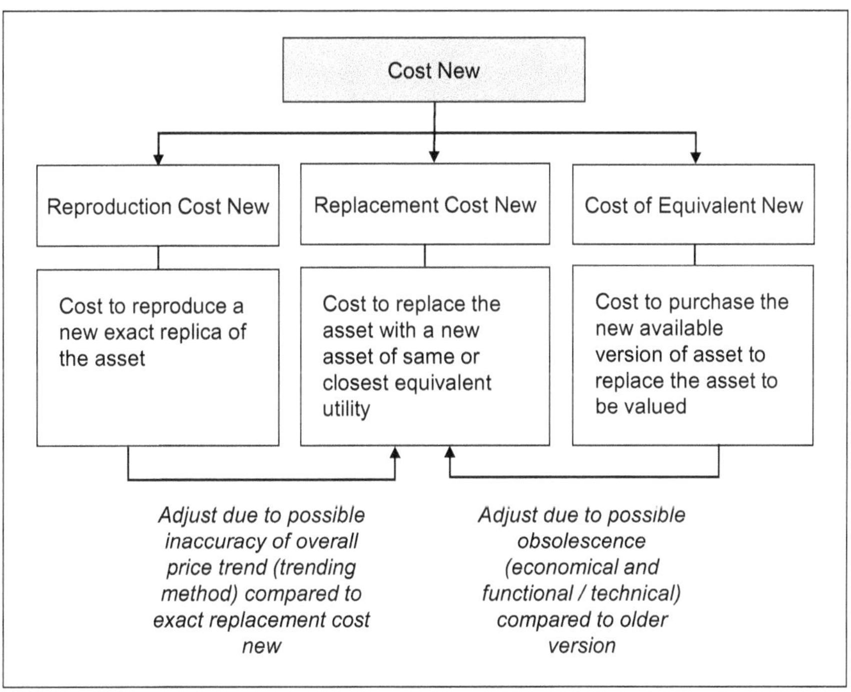

Figure 6.27 Different Types of Cost New.
Source: Authors, adapted from Plock et al. [2025], p. 13.

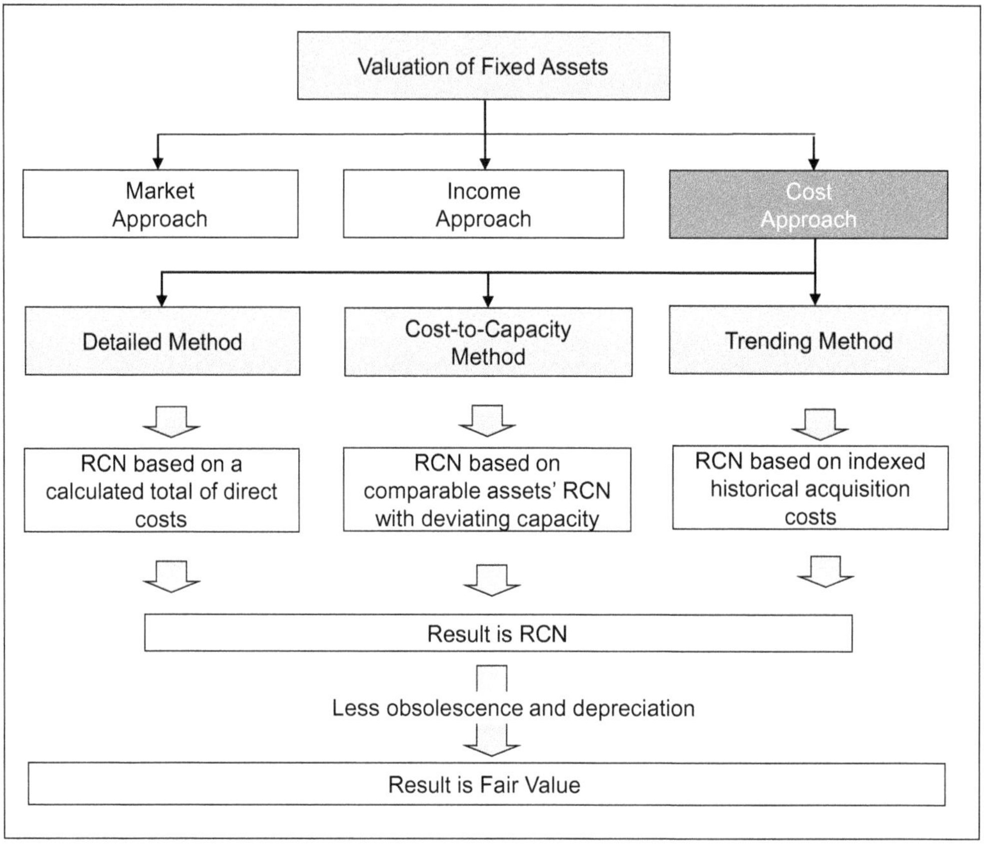

Figure 6.28 Valuation of Machinery and Equipment, Cost Approach.
Source: Authors.

improvement in value of the successor device to the total value, and the market situation for the sales side must be analyzed (see technical and economic obsolescence).

Alternatively, the value can be determined by using the reproduction cost new. The reproduction cost new describes the costs at which an asset would have to be restored to its original condition on the valuation date in a new and impeccable condition. It also includes not only the value of a machine, but also all related necessary ancillary costs, e.g. for freight, assembly, installation, foundation.[81]

Under the cost approach to derive the proper RCN there are several different methods which can be applied (Figure 6.28). The three most common valuation methods in practice are:

The Detailed Method – RCN based on a calculated total of direct costs
The Cost-to-Capacity Method – RCN based on comparable assets' RCN with deviating capacity
Trending Method – RCN based on indexed historical acquisition costs

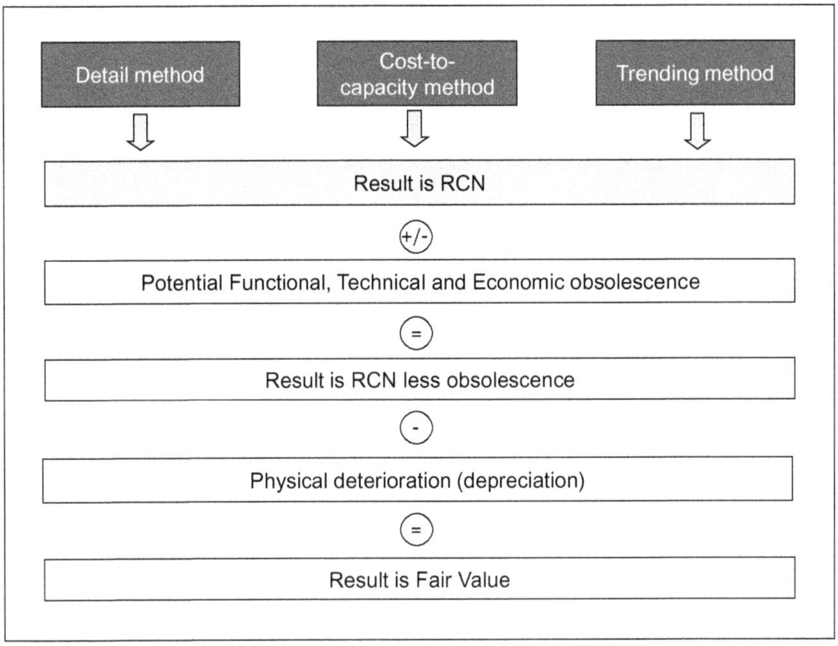

Figure 6.29 Methods to Derive RCN.
Source: Authors.
Note: Cf. Plock et al. [2025], p. 17.

The fundamental assumption of deriving the value by depreciating the respective RCN by considering the age and physical and economical condition is the same for all three methods. Hence one could argue that the different methods described are just different ways to derive the RCN representing the starting point of the cost approach.

The cost of the valuation object, as determined using one of the aforementioned methods, must now be reduced by the depreciation resulting from wear and tear, as well as technical and economic obsolescence, on the reporting date. This will be described in detail above (Figure 6.29).

A key challenge in valuation is that assets are rarely replaced with identical versions in terms of design, capacity, materials, or energy requirements. Instead, they are often substituted with more contemporary designs that better fulfill current market demands. Consequently, it is essential to make adjustments for technical and economic obsolescence to ensure that the valuation accurately reflects these changes and aligns with current market conditions and technological advancements.[82]

As illustrated in Figure 6.30 the calculated Fair Value takes into account the wear and tear over the useful life of the asset. The average useful life is determined on the basis of industry-typical empirical values for the average technical useful life of the respective class of Machinery and Equipment. In individual cases, this can deviate considerably from the depreciation period permitted for assets for tax purposes.

The calculation of the RUL may result in the assets to be valued no longer having a calculated RUL and therefore no calculated Fair Value on the valuation date. If the

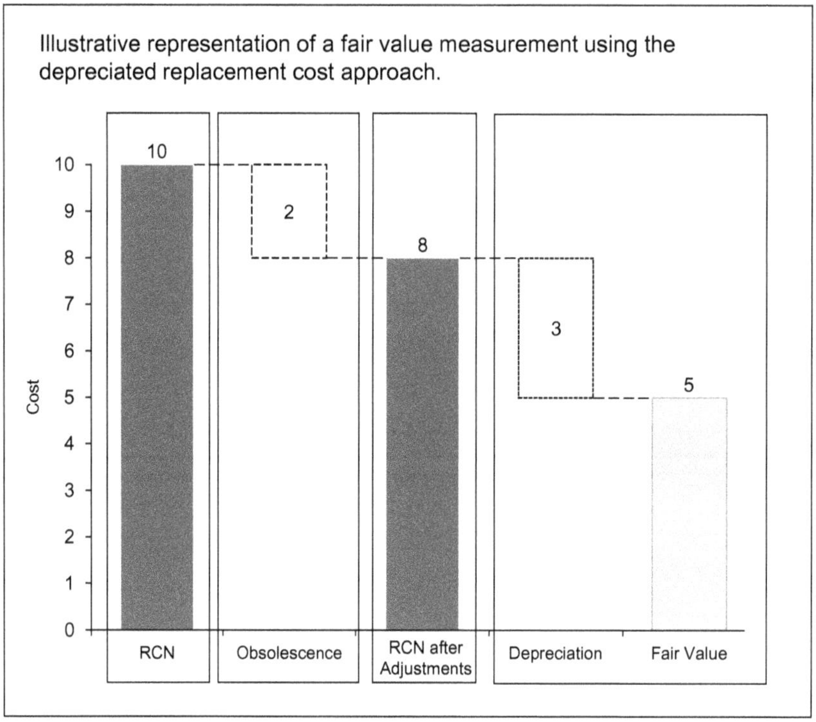

Figure 6.30 Fair Value Measurement Using DRCN.
Source: Plock et al. [2025], p. 25.

functioning asset nevertheless participates in the revenue generation for the company as part of the production process and thus generates a benefit, this circumstance is taken into account by recognizing the Residual Value, also known as the minimum operating value of the asset (Minimum Value).

The Minimum Value serves as a lower value limit. Those assets for which there is little or no calculated Fair Value due to their short calculated RUL are recognized at the higher Residual/Minimum Value if the asset is in working order.[83]

The economic current value also takes into account potential functional, technical and economic obsolescence. Technical obsolescence is caused by changes in technology, new materials and improved manufacturing processes. Economic obsolescence occurs as a result of overcapacity, changes in supply and demand or other external factors.

When performing the M&E valuation for financial reporting purposes or comparing the valuation components described above with related accounting data, it regularly can be described as shown in Figure 6.31.

Due to overall price development the derived RCN is typically higher than its historical cost of acquisition or Gross Book Value (GBV). Moreover in practice it can be observed that assets are often used longer in the production process than their applied accounting life would indicate. Hence, if no significant obsolescence has to be considered, the Fair Value derived will be larger than the asset's Net Book Value (NBV) which leads to a

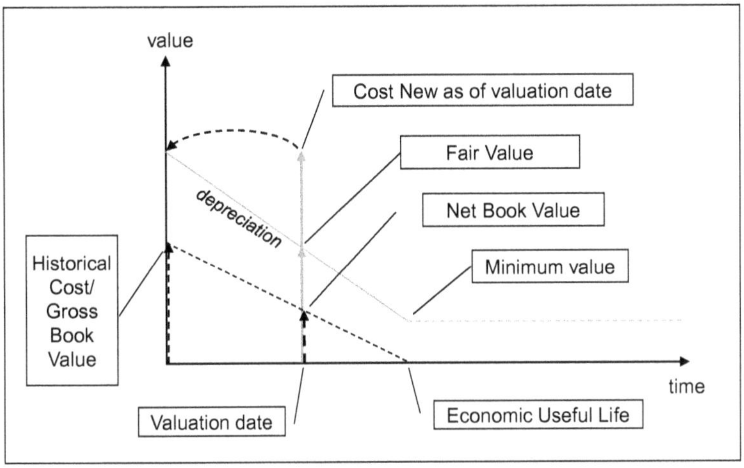

Figure 6.31 Illustration of Cost Approach.
Source: Authors, adapted from Plock et al. [2025], p. 13.

positive difference also frequently referred to as Fair Value Adjustment (FVA), if positive a "Step up". On the other hand, if the RCN are lower than historical cost or the economic useful life is shorter than the accounting life or significant obsolescence need to be applied, the derived Fair Value could be lower than its respective NBV and hence lead to a negative FVA ("Step down").

6.9.2 Detailed Method

The detailed method, also referred to as the summation method[84] is a frequently described method in the valuation literature to derive the Cost New for Machinery and Equipment assets.

Here, the valuation object is broken down into "individual parts" and researched current Cost New are assigned to each of component of the asset. These cost new include all the typical direct and indirect cost of the asset and its individual parts. These individual parts can e.g. be made up of material and labor costs and include electrical installation, foundations, freight and packaging, but also costs for development and construction, licenses and patents, financing and insurance, inspections and tests, start-up costs and interim solutions. The sum of all allowable costs results in the cost new of the valuation object.

In theory, the detailed method of cost components should provide the most precise RCN. However, its practical application is often hindered by time constraints within valuation projects or the lack of a comprehensive cost and reference library.

When valuing large individual assets, the detailed method should be applied or at least attempted, particularly if a comprehensive breakdown of all direct and indirect costs is not feasible. However, if a large quantity of assets needs to be valued, employing this method as the primary approach may not be feasible. It would be nearly impossible to determine the cost new for each individual asset in an entire factory or even a major production line in a reasonable timeframe.

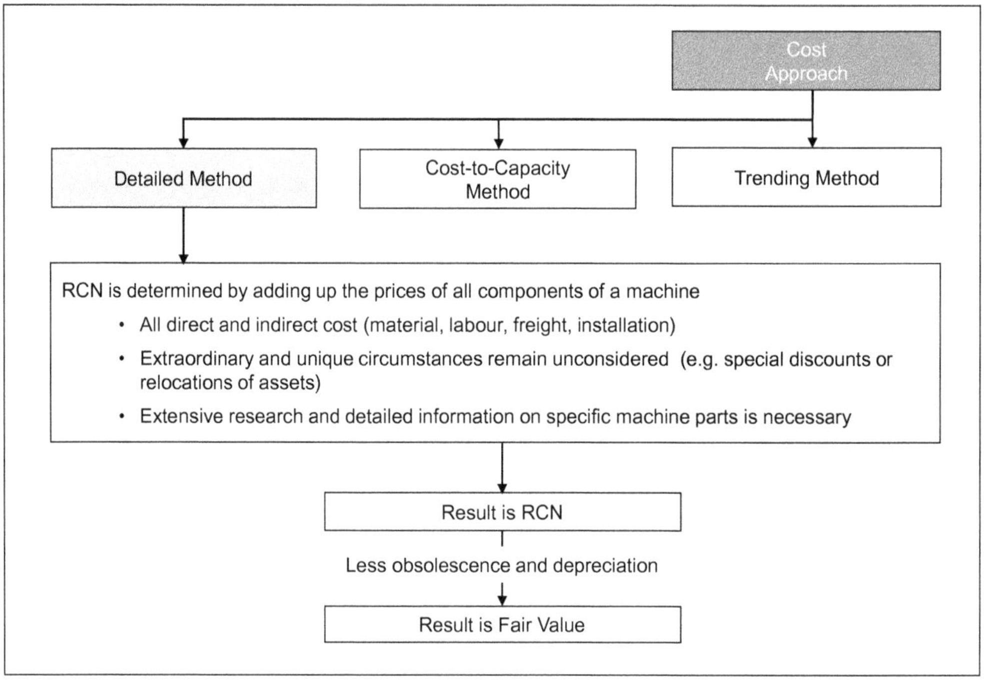

Figure 6.32 Cost Approach, Detailed Method.
Source: Authors.

Nonetheless, the detailed method proves useful and is frequently employed in valuation practice to cross-verify the accuracy of results obtained through other methods. For instance, if RCN values are determined using the trending method, the validity of these results should be checked against the detailed method at least applied to a representative sample of assets. This sample could include the most valuable assets with the greatest impact on the overall valuation or a representative sample of asset from each asset group, ensuring a broader coverage of the diverse asset portfolio.

6.9.3 Cost-to-Capacity Method

Another method to derive the RCN is the cost-to-capacity method.[85] It is often used for larger production facilities or entire plants.[86] The basis for the cost-to-capacity method is the assumption that the costs of plants of similar technology, but different sizes relate to each other but do not vary linearly but exponentially, as outlined in the equation below.[87]

By transforming the term in Cost-to-Capacity Equation (Figure 6.33), the RCN for the asset to be valued can be determined based on the known cost and capacity of the comparison asset and the known capacity of the asset to be valued.

Cost-to-Capacity relationship

$$\left(\frac{C_2}{C_1}\right) = \left(\frac{Q_2}{Q_1}\right)^X \quad \Rightarrow \quad C_2 = C_1 \left(\frac{Q_2}{Q_1}\right)^X$$

C_2 = Required cost of asset 2 (or piece of M&E 2), with capacity Q2
C_1 = Known cost of asset 1 (or piece of M&E 1), with capacity Q1
Q_2 = Known capacity of asset 2 (or piece of M&E 2)
Q_1 = Known capacity of asset 1 (or piece of M&E 1)
X = Scale factor for technology of asset 2 and 1 (or piece of M&E 2 and 1)

Figure 6.33 Cost-to-Capacity Relationship.
Source: Authors, adapted from Baumann and Lopatnikov [2017], p. 38.

A prerequisite for the application of the capacity equivalence method (Figure 6.34) is knowledge of the new value of an identical or at least similar/comparable plant in terms of technology and configuration of the plant, its location and all special design and location features.[88] The applied scaling factor should adequately reflect the technology of both the known and the estimated plants and should also be specifically applicable to the size range of the analyzed plant.[89] A frequent value often referred to in the literature for this scale factor x is 0.6 to 0.7.[90] Therefore, this relationship is often referred to as the six-tenths (0.6) factor rule.[91] Finally, the known historical costs must also be adjusted to the valuation date in order to take into account any price developments and, if applicable, exchange rate changes.

The cost to capacity method is regularly applied to derive RCN for large assets such as entire production plants, power plants or large petro-chemical facilities. For example, when a rough ballpark figure for construction cost of an entire petrol chemical factory or power plant needs to be estimated and the applied technology as well as capacities is known, the cost to capacity method might be a helpful proxy to derive the overall RCN of such facilities.

However, when a large difference in capacity is being calculated the described cost-to-capacity rule should be considered with caution and derived results crosschecked by their plausibility.[92]

Moreover, in practice the cost-to-capacity approach has similar challenges as described earlier for the detailed method. When valuing a large quantity of single assets, it will not be applicable in practice since it is too time-consuming and the required data may not always be available. In such case the Trending Method is applied.

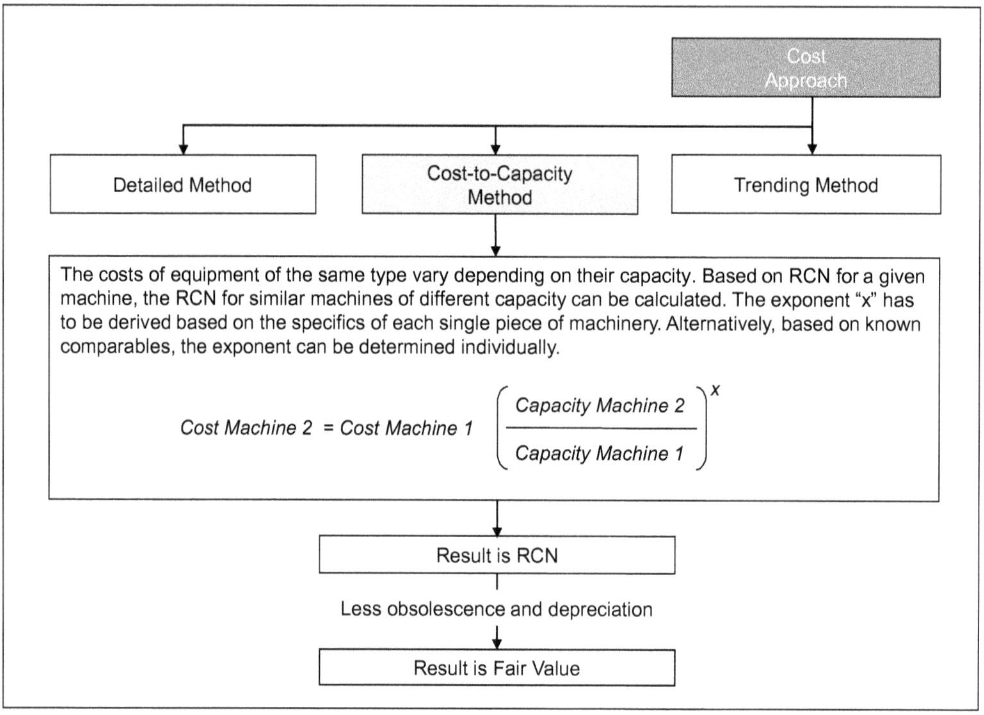

Figure 6.34 Cost Approach, Cost-to-Capacity.
Source: Authors.

6.9.4 Trending Method

The Trending Method also referred to as Index Method[93] is the most frequently used method to determine the cost new of an asset as of valuation date.[94] This method involves applying an index that reflects the price changes for specific assets or asset categories to convert the historical cost of acquisition or production into an estimate of current cost (Figure 6.35). It is crucial to differentiate between historical cost and original cost, as they may not always be identical. According to the ASA, historical cost is defined as "the cost of an asset when it was first placed into service by its first owner."[95]

In contrast, original cost, often referred to in accounting as Gross Book Value, is defined as "The initial capitalized cost of an asset in the hand of its present owner who may not be the first owner and who may have purchased at a price greater or less than the historical cost – may be the used cost of the property."[96]

To ensure the trending method is accurately applied, the valuer must confirm that the cost basis used is the historical cost of the asset, reflecting the market level at the time of acquisition. Historical acquisition or production costs include all expenses incurred to acquire and initially operate the Machinery and Equipment. The net invoice value serves as the basis for this calculation, excluding discounts and special conditions. In cases where company records are unavailable, supplier information and price lists may be used to estimate these costs.[97]

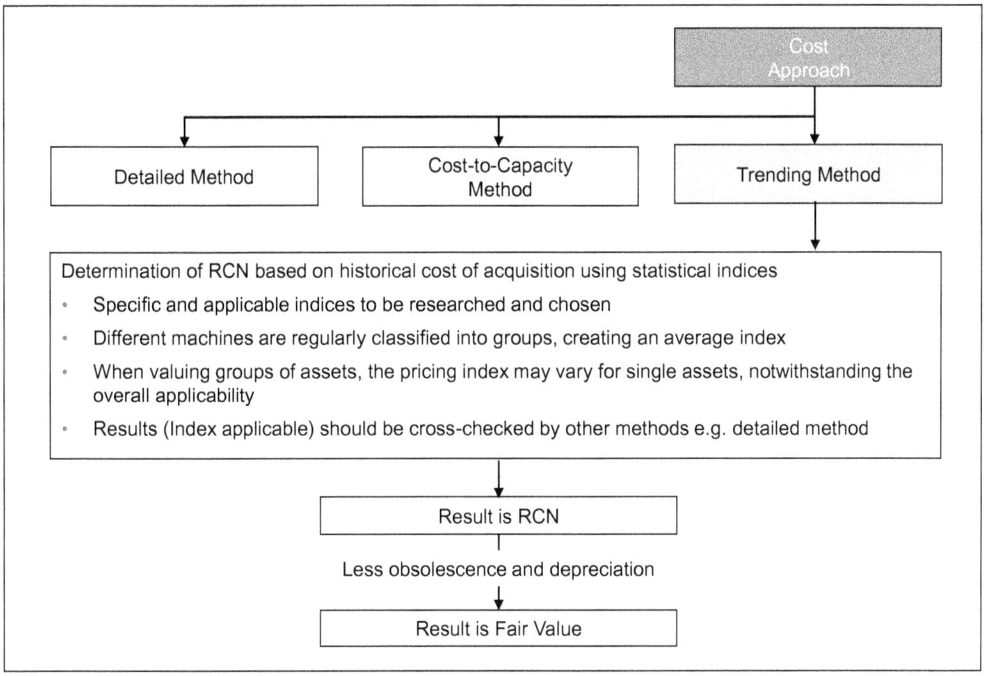

Figure 6.35 Cost Approach, Trending Method.
Source: Authors.

In the Trending Method, an index is utilized to calculate the Cost New by adjusting the historical cost to account for price changes over time. As shown in Figure 6.36, this involves dividing the current index of valuation date by the initial or base value (set at 100) of the time series, allowing the index numbers to be interpreted as a percentage of the initial value. The sample table below illustrates how this calculation is applied, with each index number representing the proportionate change from the base value. This approach provides a clear and systematic way to estimate the current replacement cost by reflecting historical cost adjustments in line with market trends.[98]

Besides the historical cost the consideration of a correct index, meaning an index that correctly reflects the individual assets price development since its acquisition is another precondition for a proper application of the Trending Method.

Indices are typically sourced from national statistical offices, e.g. Eurostat for the European region, or other authoritative bodies, research organizations, or service providers. When employing the index method, it is crucial to select the most appropriate index for the specific asset type.[99]

When utilizing price indices, it is crucial to understand that they reflect a genuine price change over a specific period. Consequently, when establishing these indices, it is ensured that variations in product specifications (such as quality and features) and sales conditions are excluded from the index calculations.

When selecting the index, care must be taken to ensure that it corresponds to the actual historical price development of the asset to be valued. In individual cases, it will often

Year	Price	Index
2016	99	100
2017	105	106
2018	110	111
2019	115	116
2020	119	120
2021	119	120
2022	123	124
2023	138	139
2024	145	146
2025	151	153

Formula to derive RCN as of valuation date:

$$\left(\frac{\text{Current Index}}{\text{Base Year Index}}\right) \times \text{Historical Cost} = \text{RCN}$$

$$\left(\frac{138}{99}\right) \times \text{Historical Cost (2016)} = \text{RCN (2023)}$$

Figure 6.36 Price Index.
Source: Authors.

not be possible to find an individual index for all specific machine or plant to be valued. In such cases, the next higher level of the machine group must be used. This is usually sufficiently accurate and acceptable for the valuation. However, care must be taken not to use "global" price indices that also include price developments of other economic goods and therefore do not reflect the price development of the individual assets to be valued. Global price indices that reflect the developments of an entire economy are not sufficiently accurate and are therefore not applicable. Additionally, care should be taken to ensure that not only are the historical acquisition or production costs properly indexed, but also that not all costs recorded in a contract may align with a particular index.

Therefore a critical plausibility check of the results is recommended, especially with long-term index series, to ensure the applied index accurately reflects price developments also for those older assets.

When appraising international assets or asset portfolios, it is important to recognize that using index series from other countries may yield inaccurate results. This is particularly true in the context of significant exchange rate fluctuations, necessitating a thorough assessment of the data to ensure reliability.

To manage a larger volume of assets during valuation, the Fixed Asset Register serves as the foundational tool for the cost approach and the trending method in particular. As discussed in Chapter 5, the fixed register is essential for implementing the trending method effectively.

6.9.4.1 Historical Acquisition and Production Costs

The concept of historical cost or Historical Cost of Acquisition (HAC) is rooted in corporate accounting practices. IFRS provide guidance on the acquisition and production costs for property, plant, and equipment in IAS 16.[100]

For an accurate valuation, it is essential to use the original acquisition and production costs from when the asset was first put into service by its initial user. These costs may sometimes differ significantly from those incurred by the current owner.[101]

Regularly obtaining acquisition cost information from the asset accounting system, such as the Fixed Asset Register, is crucial. It is important to understand the framework underlying the Fixed Assets Register e.g. whether this register is maintained according to commercial law or tax law, as differences in asset capitalization can arise between the two.

In practice, the issue of second-hand assets, which involves distinguishing between historical cost and original cost, can arise, particularly when assets are acquired not individually on the market but through company acquisitions or mergers with another company. In such instances, it is critical to verify whether the original cost of acquisition recorded in the asset register accurately reflects the initial acquisition prices. The same scrutiny should be applied to the asset's age. When assets have been acquired second-hand or as part of an entire company acquisitions, their actual age may differ from what is indicated by the acquisition or access date in the asset accounting records.

When carrying out the on-site inspection, it is advisable to verify the year of manufacture on the basis of the machine type plate.

To address these discrepancies, invoices should be re-examined at least for a reasonable sample of representative assets to confirm the exact acquisition date and price. In this context it is crucial to ensure that the historical cost reflects market prices as of the historical acquisition date. The trending method relies on the assumption that the original acquisition costs are aligned with the Market Value. Only under this condition can the indexed costs be used to estimate the Market Value accurately. If there were any substantial discounts given by the seller which are individually obtained only by this specific buyer and not granted to any third party, the historical acquisition cost doesn't reflect market conditions. Only if the HAC reflects market prices as of acquisition date can the trended historical cost reflect market price and Market Value as of valuation date.

Another factor that needs to be considered when assessing the historical acquisition cost for the trending method is that HAC can consist of different components.

If the asset's purchase price paid by the owner significantly deviates from what an unrelated market participant would have paid, these acquisition costs should not be used as a basis for valuation. In this context a distinction must be made between direct and indirect acquisition costs such as packaging, shipping, insurance, commissioning, etc., other components such as foundations, ventilation or electrical systems, etc., as well as a valuation of own and third-party services and, if applicable, intangible assets such as research and development costs, patents, etc.[102]

6.9.4.2 Strength and Weaknesses of the Method to Derive the RCN

The three methods have advantages and disadvantages (see Figure 6.37) which should be considered when choosing any of them.

The detailed method, also known as the direct approach, is recognized as the most precise technique for determining the RCN of an asset. This approach directly correlates with market developments, reflecting the price a market participant would be willing to pay for a new asset as of the valuation date. However, in practice, it is quite time-consuming to research detailed RCN for each asset, especially when dealing with large volumes or specialized assets. This challenge is akin to the issues faced with the market approach, as identifying costs can be particularly difficult for specialized or custom-built assets. Additionally, when valuations are conducted for financial reporting purposes,

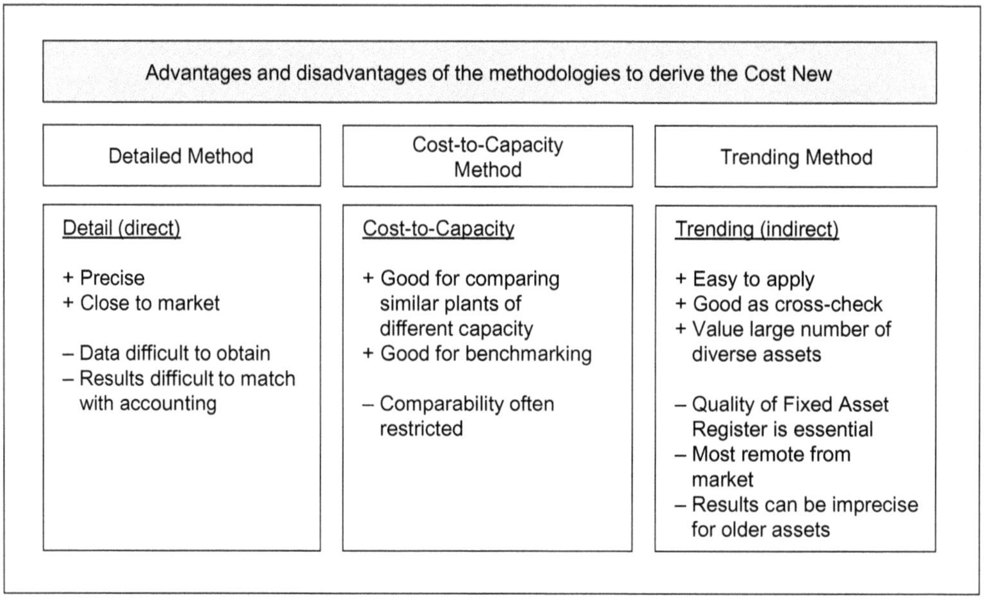

Figure 6.37 Advantages and Disadvantages of Methodologies to Derive Cost New.
Source: Authors adapted from Plock et al. [2025], pp. 26–28.

aligning the results with accounting figures can be challenging, if the accounting data is structured differently.

The cost-to-capacity method is often used to compare similar assets or plants of different capacities, and it is a practical method for benchmarking or cross-checking results obtained from other approaches. However, its comparability can be limited in practice, which poses a disadvantage.

The trending method is the most commonly applied technique for valuing Machinery and Equipment assets, particularly when the valuation involves a large number of assets. This method is transparently structured, straightforward and efficient for managing substantial asset volumes. However, it relies on certain prerequisites, such as a reliable Fixed Asset Register with robust historical acquisition costs (reflecting market conditions at acquisition date) and a specific price index to derive the RCN. Critics point out that this approach is somewhat detached from current market developments, potentially leading to imprecise valuations, especially for older assets.

Therefore, conducting a thorough cross-check of the derived RCN using other methods like the detailed method or the cost-to-capacity method is essential, at least for a sample of major assets, to ensure accuracy and reliability in the valuation process.

6.9.5 Obsolescence

When current indexed prices are used to derive the RCN, potential allowance must be made for advances in production technology. Where an asset is replaced by one of the same type and capacity which is to be used for the same purpose, advances in technology also may necessitate deductions for technical changes (obsolescence) which will depend

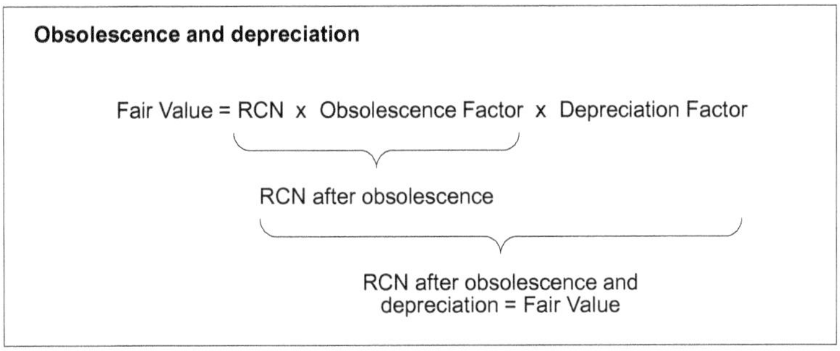

Figure 6.38 Obsolescence and Depreciation.
Source: Authors.

on the proportion of the replacement asset's total value represented by the increase in value resulting from the changes (Figure 6.38).

Generally, it is distinguished between economic obsolescence, functional obsolescence, and technical obsolescence.

In practice it can be observed that sometime the obsolescence is considered after depreciation for physical deterioration. Since both steps – the consideration of obsolescence as well as the depreciation – are mathematically multiplications of the RCN with certain factors, the order does not make a difference.

Three different types of obsolescence are increasingly being identified in practice (Figure 6.39).[103]

Economic obsolescence: This refers to the loss in value of the machine due to external influences.
Functional obsolescence: Refers to the loss in value of the machine due to the existence of more efficient, comparable machines.
Technical obsolescence: Refers to the loss in value of the asset due to the machine becoming unusable.[104]

6.9.5.1 Economic Obsolescence

Economic obsolescence arises from external factors that affect a property's or asset's value, often due to overcapacity or shifts in supply and demand.[105] Key contributors include the overall economic condition of the industry, availability of financial resources, loss of materials or labor, enactment of new laws or regulatory changes, rising costs of raw materials, labor, or supplies without a corresponding increase in product prices, reduced demand for the product, heightened competition, inflation, high interest rates, and similar factors.[106]

Economic obsolescence occurs as a result of overcapacity, changes in supply and demand, or other external factors.[107] The most common reasons for this are[108]:

- The economic situation of the industry
- The availability of financial resources
- The loss of material and/or labor
- The adoption of new laws or changes to regulations

- Increased raw material, labor or supply costs (without compensatory increases in product prices)
- Decreased product demand

6.9.5.2 *Functional Obsolescence*

Functional obsolescence refers to the loss of value or usefulness of an object caused by its uneconomical nature or inadequacies in comparison to a more efficient or cost-effective comparable object developed using new, modern technologies.[109] It is usually caused by technological changes, new materials and improved manufacturing processes.[110] Corresponding characteristics are[111]:

- Increased operating costs
- Higher space requirements
- Efficiency in production (e.g. maximum speed)
- Lack of flexibility in adapting to new products
- ESG criteria, such as CO_2 emissions
- Work quality measured by the deviation tolerance (e.g. thread diameter on lathes)[112]

6.9.5.3 *Technical Obsolescence*

Technical obsolescence refers to a loss in value caused by wear and tear, ageing, and a lack of maintenance, which impairs the physical condition of the machine. This reflects the extent to which the asset's physical functionality and efficiency has deteriorated over time. Technical obsolescence occurs when machines participate in the production process for significantly longer than their technical service life allows. Reasons are:

- Wear and tear
- Material fatigue of the machine
- Maintenance condition is influenced by the level of maintenance measures carried out
- Age and service life
- Environmental influences, e.g. corrosion[113]

As shown in Figure 6.39, the RCN needs to be adjusted by all potential obsolescence factors (if needed: economical, technical and functional) before the adjusted RCN (RCN after adjustment) is depreciated.

6.9.6 *Depreciation*

The replacement cost (new) of assets is adjusted for the loss in value caused by physical deterioration. Depreciation for valuation purposes is the estimated loss in value of an asset, compared with a new asset; depreciation measures value inferiority that is caused by physical deterioration, functional obsolescence, and economic obsolescence.[114]

The arithmetical Fair Value is calculated from the cost new, taking into account age-related depreciation, or from the acquisition value of the asset, taking into account the change in price conditions. In general, there are three methods employed for this purpose[115] (Figure 6.40):

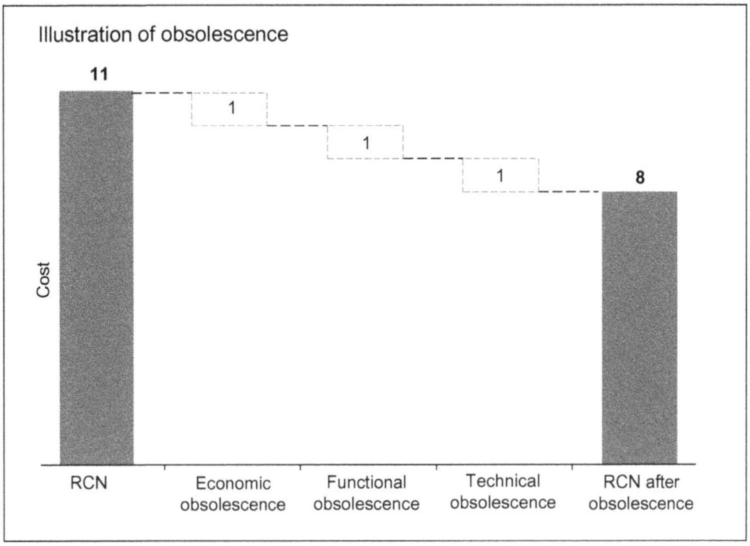

Figure 6.39 Illustration of Obsolescence.
Source: Plock et al. [2025], p. 43.

Straight-line depreciation: The asset's value decreases by the same amount each year.[116]
Progressive depreciation: The asset is used less in the early years, but more over time.
Declining balance depreciation: The asset loses a significant amount of its value in the first few years, with the rate of depreciation levelling off over time.[117]

Different methods of calculating depreciation can have a significant impact, particularly in the early years. The depreciation of Machinery and Equipment is usually progressive. In practice, however, linear depreciation is usually used instead, which implies a risk discount, especially in the first few years.[118]

The last two methods (arithmetical/geometrical decline) are based on the premise that the value of an asset falls more quickly in the first few years of its life than in later years and are thus more likely to reflect the true loss in value of assets that are not very old. For plant and machinery comprising a high proportion of long-lasting assets sufficiently accurate results can be obtained by using the straight-line method.[119]

6.9.6.1 Economical and Technical Useful Life

The average useful life of an asset is estimated on the basis of empirical industry data on typical average technical useful lives of assets of the category concerned. It is the estimated number of years that a new property will actually be used before it is retired from service. A property's normal useful life relates to how long similar properties actually tend to be used, as opposed to the more theoretical economic life calculation of how long a property can profitably be used. The useful life of an asset also substantially depends on its use and maintenance.[120]

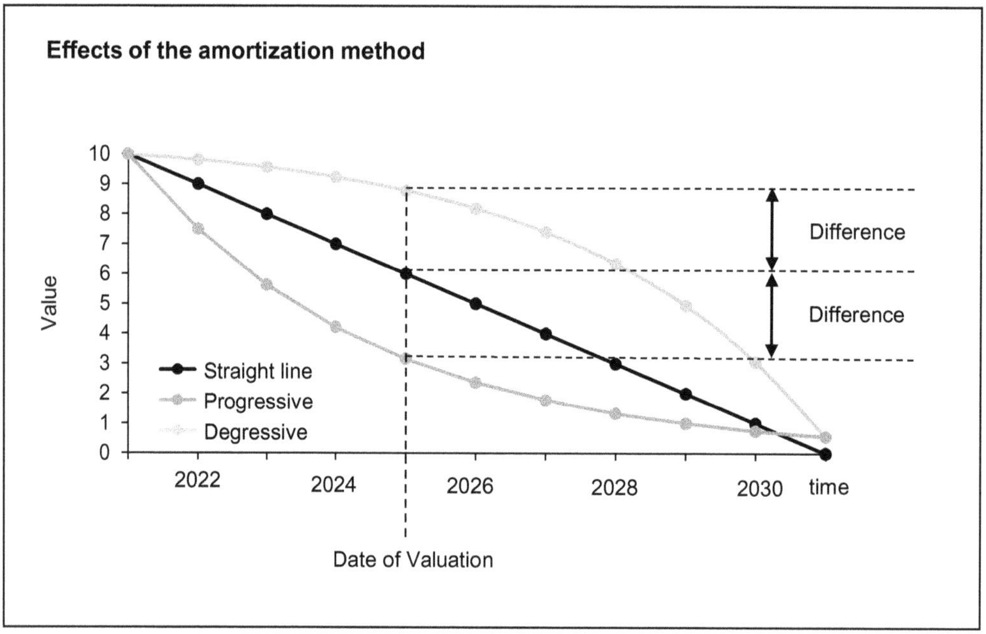

Figure 6.40 Effects of the Amortization Method.
Source: Plock et al. [2025], p. 63.

The average useful life may differ considerably from the depreciation period allowed for tax purposes. The RUL is calculated from the average useful life and the year of acquisition.[121]

Technical useful life: This is the period during which an asset remains functional under normal operating conditions. It takes into account the physical wear and tear, as well as the natural deterioration, of an object. The duration is influenced by construction, materials, and maintenance levels.[122]

Economic useful life: This refers to the period during which an asset can be used in an economically meaningful way. This means that during this period, the asset generates economic benefits or profits that justify the costs of acquisition, operation and maintenance. This useful life can be influenced by market conditions, technological developments or changes in the law. This can result in the asset becoming unprofitable before the end of its technical useful life.[123]

The economic life is not always identical to technical life. Rather, technological progress, legal requirements, the costs associated with operation, or the availability of spare parts must be taken into account when determining the economic useful life.

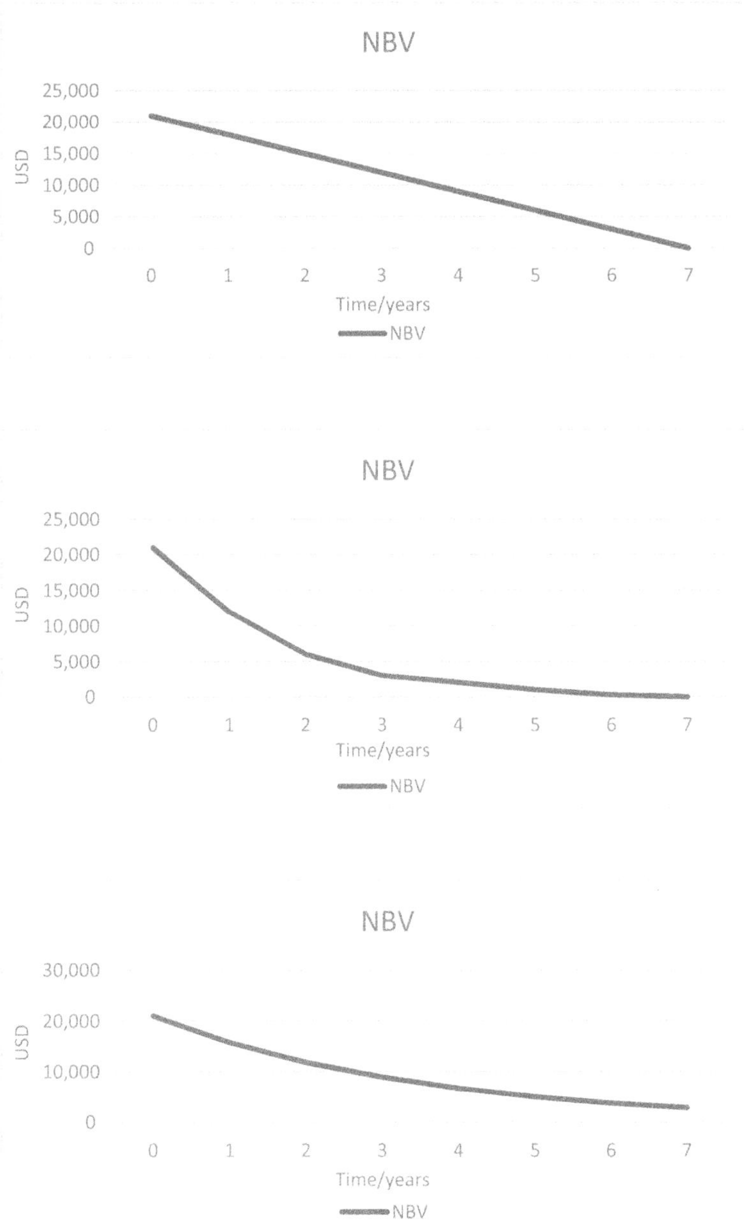

Figure 6.41 Physical Deterioration Based on Age.
Source: Authors according to Plock et al. [2025], p. 17.
Notes: Cf. IfS [1999], p. 18.

Calculating average weighted age

Year of acquistion	GBV		Trendfactor		RCN	Age*	Weighted RCN
2015	11,900	x	2.6	=	30,940	10	309,400
2018	2,000	x	1.95	=	3,900	7	27,300
2020	2,500	x	1.6	=	4,000	5	20,000
2023	17,500	x	1.2	=	21,000	2	42,000
Total	**33,900**				**59,840**		**398,700**

*Date of valuation 2025

$$\text{Average weighted age: } \frac{398{,}700}{59{,}840} = \mathbf{6.7\ years}$$

Figure 6.42 Calculating Average Weighted Age.
Source: Authors.

Depending on the definition of the valuation object, components with different useful lives are included. As a rule, short-lived components can be replaced without jeopardizing the use of the valuation object. As part of the valuation, the valuer must ensure that the availability and thus fulfilment of the function of short-lived components is guaranteed and that there is no risk to the economic use over the assumed duration.

This risk has increased due to increasingly rapid technological development, legal requirements (e.g. environmental regulations, inspection obligations) or the availability of the components themselves.

When analyzing industry-typical empirical values, particular attention must be paid to the history of use in the underlying valuations. Issues such as maintenance management (preventive or breakdown-orientated maintenance), replacement of components, overhauls, damage and repairs, etc. play a key role here.

The RUL is calculated from the year of acquisition of the asset and the average useful life.[124]

This cut-off value may represent the Minimum Value assigned to the asset in the valuation study. It is attributed to functioning assets possessing a lower or no arithmetical value because they have little or no RUL.[125]

6.9.6.2 Remaining Useful Life

The RUL is normally the difference between the expected TUL of an asset and its age at the valuation date.[126] It must then be considered whether the arithmetically determined RUL can be achieved, in light of the asset's condition, the general and local economic

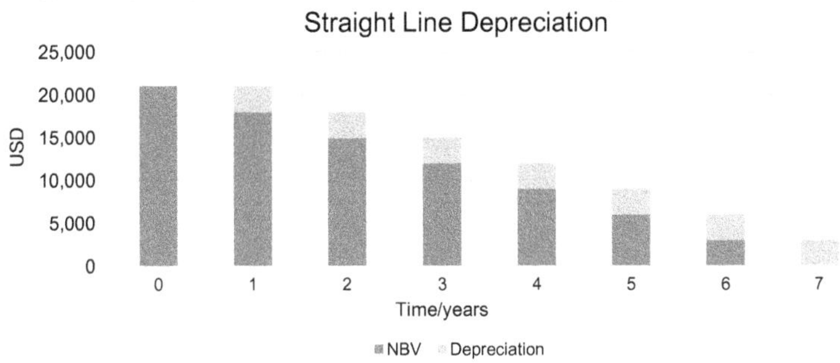

Figure 6.43 Straight Line Depreciation.
Source: Authors.

situation and the asset's future usability (the demand for its use). If not, the RUL must be reduced accordingly.[127]

The RUL life is thus the period, commencing with the valuation date, during which the building can be expected to be capable of use for economic purposes. As a rule, it is shorter than the asset's technical life, which depends largely on the construction of the foundations and other load-bearing components.[128]

6.9.6.3 *Physical Deterioration*

Physical deterioration is the loss in value or usefulness of a property due to consuming its useful life caused by wear and tear, deterioration, exposure to various elements, physical stresses, and similar factors.[129]

After deducting all three potential forms of obsolescence (economic, functional and technical if required) and depreciation due to physical deterioration the resulting depreciated replacement costs represent an estimate for Fair Value of the asset according to the cost approach.

6.9.6.4 *Application of Minimum Value*

When calculating the RUL, it is possible that Fixed Assets valued on the valuation date will have no RUL, and therefore no Fair Value, calculated. However, if the asset generates revenue for the company as part of the production process, providing a benefit, this is taken into account by applying the Minimum Value.[130]

The Minimum Value therefore serves as a lower limit. Assets for which a low or zero Fair Value can only be determined due to their short RUL are assigned a higher replacement value if they are still functional.

As defined in Chapter 1, the Minimum Value is often applied at 5% or 10% of the replacement or reproduction cost of new assets, depending on the industry.

6.9.6.5 Summary and Practical Application of the Cost Approach

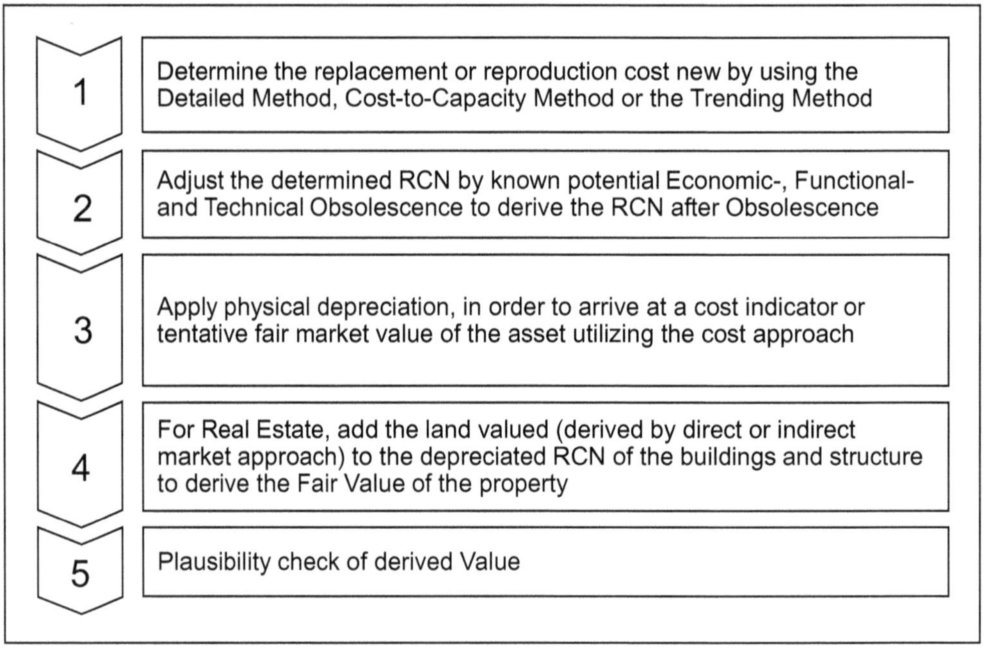

Figure 6.44 Practical Application of the Cost Approach.
Source: Authors.
Note: Cf. ASA [2011], p. 81.

6.10 Plausibility Check

A thorough valuation process includes a plausibility check and categorization of the valuation outcome. Ideally, at least one alternative valuation method should be employed, and any discrepancies between results should be thoroughly analyzed and justified.[131]

For plausibility checks, resources such as list prices, used machinery platforms from dealers, and trading platforms for pre-owned goods can be utilized. It is crucial to pay special attention to factors such as usage history, the current condition of the asset, and any overhauls or refurbishments.

Additionally, comparisons should incorporate considerations of the relevant market, warranties, ancillary acquisition costs, and any additional work required. This comprehensive approach ensures a more accurate and reliable valuation by accounting for all pertinent factors.

6.11 WARA Test

In order to assess the reasonability or plausibility of the value derived for the different Fixed Assets of a company, the so-called Weighted Average Return on Assets (WARA) test can be applied.

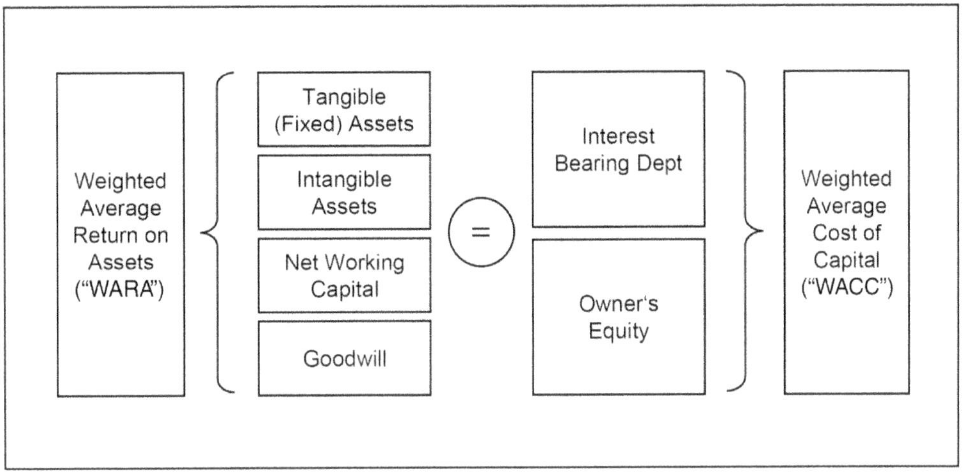

Figure 6.45 WARA Test.
Sources: Crane [2018], p. 12 and Zyla [2013].

The WARA test is particularly used in practice in the context of financial reporting and during purchase price allocations (PPAs) for business combinations. The WARA is a corroborative tool used to assess the reasonableness of the Fair Values assigned to acquired assets, especially intangible assets like brands and customer relationships. It essentially checks if the collective rates of return on the various assets, weighted by their respective values, align with the overall cost of capital (WACC) for the acquired business. The WACC represents the overall return required by both debt and equity holders.[132]

The process begins with the calculation of WARA, which represents the average expected return on a company's assets. This is done by weighing the return on each asset class according to its share of the total assets. Besides Fixed Assets the calculation considers asset categories such as working capital, and intangible assets.[133]

Once the WARA is determined, it is compared with the company's WACC. Ideally, the WARA should be consistent with or slightly lower than the WACC, reflecting the risk profile of the assets. If the WARA is significantly higher than the WACC, it may suggest that the asset values or expected returns are overstated. Potential benchmarks of a typical financial reporting valuation for an industrial company for the various asset classes are: Intangible assets 8%, Tangible Assets 6%, Net Working Capital 5%, and Goodwill 10%. Those figures can't be recognized as absolute benchmarks but show the relation of the asset's risks among each other.

By ensuring consistency in asset valuations with the overall company value and return expectations, the WARA test promotes transparency and accuracy in financial reporting.

6.12 Valuation of Work in Progress Assets

Work in progress (WIP) or work under construction refers to assets that are not yet completed or fully operational. Under the going concern assumption, the value of these assets is typically recognized at their NBV, which reflects the costs incurred up to that point. The rationale behind this approach is that if the asset was acquired at market price, the NBV should reasonably represent its Fair Value.

However, in practice, it is crucial to verify that the NBV accurately reflects market prices. Depending on the volume of WIP assets and their proportion of the company's overall Fixed Assets, it is advisable to cross-check a reasonable sample to ensure that the recorded prices truly align with Market Values and that the assets are indeed new. Additionally, it is important to determine how long the assets have been classified as WIP.

During the valuation process, assessing the progress status of these assets is essential, as well as valuing whether they are genuinely needed for the business' operational processes. This ensures that the valuation accurately reflects the current status and necessity of the WIP assets within the company.

6.13 Derivation of Fair Value Adjustment

If the valuation has been carried out for financial reporting purposes e.g. PPA, the derived Fair Value needs to be allocated to the purchase price paid for the acquired company or entity.

As shown in the illustrative examples of Figures 6.47 and 6.48 the difference of the Fair Value and its respective NBV is calculated as the FVA.

In USDm **Entity**	**Asset Gro**	**GBV**	**Depr.**	**NBV**	**FV**	**FV Adj.**
	Land	0.25	0.0	0.25	0.28	0.3
Company A	B&S	9.4	2.2	7.2	7.9	0.7
	M&E	9.0	2.0	7.0	7.5	0.5
	Subtotal	*20.9*	*4.2*	*16.7*	*18.2*	*1.5*
	Land	1.4	0.0	1.4	2.6	1.2
Company B	B&S	5.5	2.5	3.0	4.1	1.1
	M&E	8.4	5.2	3.2	5.2	2.0
	Subtotal	*15.3*	*7.7*	*7.6*	*11.9*	*4.3*
	Land	12.5	0.0	12.5	19.8	7.3
Company C	B&S	35.6	21.0	14.6	19.5	4.9
	M&E	45.0	37.5	7.5	18.9	11.4
	Subtotal	*93.1*	*58.5*	*34.6*	*58.2*	*23.6*
	Land	0.0	0.0	0.0	0.0	0.0
Company D	B&S	0.0	0.0	0.0	0.0	0.0
	M&E	1.2	1.0	0.2	0.2	0.0
	Subtotal	*1.2*	*1.0*	*0.2*	*0.2*	*0.0*
Total Group		130.5	71.4	59.1	88.5	29.4

Figure 6.46 Derivation of Fair Value Adjustment.
Source: Authors.

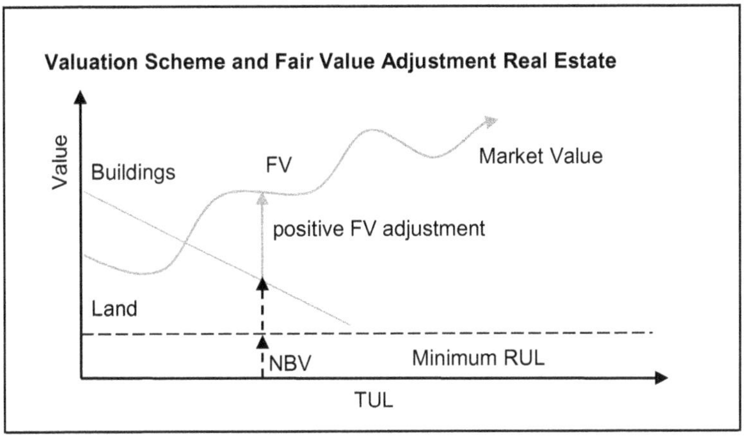

Figure 6.47 Derivation of Fair Value Adjustment for Real Estate.
Source: Authors.

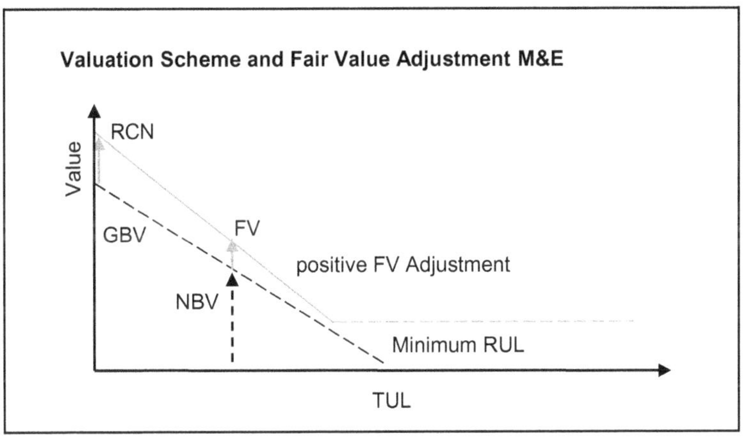

Figure 6.48 Derivation of Fair Value Adjustment for M&E.
Source: Authors.

The FVA of an asset can change over time as its respective Fair Value and NBV changes as well. Hence as the valuation result itself, it can be only derived for a spe cific date.

6.14 Regression Analysis

As described in Chapter 4 in the identification phase, when large asset portfolios have to be valued, they are divided into different clusters. A typical structure is that Cluster A, which comprises the majority of assets, is valued using the full scope valuation approach. Cluster B is often valued using the limited scope approach, meaning that a detailed valuation is performed on an asset level, but no on-site inspection is carried out and thus the assets are valued using a desktop approach. Cluster C usually accounts for only a very small portion of the total fixed asset portfolio, and approximate analysis approaches can sometimes be used for this cluster.

One such approach is regression analysis. The approach is based on the idea to utilize the valuation results derived in a detailed valuation exercise for the assets of Cluster A and Cluster B and using them to derive an indication for the Fair Value of Cluster C.

Therefore, the ratio of each asset's GBV to NBV and NBV to FV of Cluster A and B is determined and the result applied on the GBV/NBV ratio of Cluster C to derive the indicative Fair Value.

It has to be highlighted that this approach is a high-level analysis only and may not be suitable for any valuation occasion. However, if the asset structure to be valued is homogeneous and a regression analysis of the derived values for Cluster A and B can be successfully done with reasonable significance, in practice, this approach may serve as a helpful approach to derive a reasonable indicative Fair Value for minor Fixed Assets included in Cluster C.

Notes

1 Cf. Sehab [2025], online.
2 Cf. Sehab [2025], online.
3 Cf. TEGOVA [2022], p. 320.
4 Cf. Fahrländer and Kloess [2024a], p. 331; Kleiber [2018], p. 613.
5 TEGOVA [2022], p. 320.
6 Cf. TEGOVA [2022], pp. 96ff.
7 Damodaran [2005a], p. 1.
8 Also see Kleiber [2020], p. 1315.
9 The fact that the asset value method occupies such a dominant position in the valuation of property, plant and equipment is certainly a specialty. When valuing other items such as companies, property or intangible assets, however, the asset value method tends to be the exception. Here, the comparative value method and in particular the capitalized earnings value method plays the central role due to the closer market reference. One of the reasons for this is that the cash flows are easier to derive and can be better allocated to the assets.
10 Cf. Salden [2014], p. 85.
11 KPMG [2024], p. 62.
12 Cf. IFRS13, Paragraph 72–90.
13 KPMG [2024], p. 63.
14 Wyatt [2013], p. 115.
15 Cf. RICS [2007a].
16 White et al. [2003], p. 94.
17 Cf. ASA [2021], p. 114.
18 Cf. ASA [2021], p. 87.
19 Cf. Leopoldsberger et al. [2016], p. 459.
20 Cf. Wyatt [2007], pp. 110ff.
21 Cf. The Dictionary of Real Estate Appraisal, Third Edition; Appraisal Institute, 1993.
22 Cf. Wyatt [2007], pp. 110ff.
23 Cf. Corgel et al. [2001], p. 308.
24 White et al. [2003], p. 85.
25 Hackelberg and Hennig [2021], p. 328.
26 Zimmermann [2010], p. 221.
27 Cf. Corgel et al. [2001], p. 308.
28 IDW S10 [2013], p. 12.

29 Cf. ASA [2011], p. 98.
30 Corgel et al. [2001], p. 321.
31 White et al. [2003], p. 95.
32 Cf. Isaac and O'Leary [2013], p. 17.
33 Cf. Brealey and Myers [2011], p. 21.
34 Cf. Hackelberg and Chan [2021], p. 108.
35 Cf. Wyatt Peter [2007], p. 130.
36 Cf. Hagen and Reif [2004], p. 43.
37 Cf. Metzger [2018], p. 176.
38 Cf. Leopoldsberger et al. [2016], p. 465.
39 Cf. Damodaran [2005a].
40 Cf. IDW S 10 [2013], No. 3.1.3.2.
41 Cf. Hackelberg and Hennig [2021], p. 330.
42 Cf. Isaac and O'Leary [2013], p. 18.
43 Cf. Leopoldsberger et al. [2016], p. 461.
44 Also see ASA [2021], p. 125.
45 Cf. Wöhle [2005], p. 219.
46 Corgel et al. [2001], p. 324.
47 Isaac and O'Leary [2013], p. 57.
48 Fahrländer and Kloess [2024a], p. 359.
49 Isaac and O'Leary [2013], pp. 73ff.
50 White et al. [2003], p. 98; Cf. IDW S 10 [2013], para. 2.3.2.
51 Cf. Renner and Sohni [2012], p. 229.
52 Cf. White et al. [2003], p. 108; Cf. Baum et al. [2018], pp. 66ff.
53 Cf. Salden [2014], p. 79.
54 Cf. Leopoldsberger et al. [2016], p. 471; Cf. Isaac and O'Leary [2013], p. 28.
55 Cf. Isaac and O'Leary [2013], p. 28.
56 Cf. Renner and Sohni [2012], p. 229.
57 White et al. [2003], p. 112.
58 Also see: Renner and Sohni [2012], p. 229.
59 Cf. Salden [2014], p. 74; IDW S 10 [2013], para. 3.2.5.
60 Cf. White et al. [2003], p. 121.
61 Cf. Renner and Sohni [2012], p. 240.
62 Cf. Wyatt [2013], p. 187.
63 Cf. Leopoldsberger et al. [2016], p. 477.
64 Cf. Salden [2014], p. 77.
65 Cf. Kleiber [2018], p. 1273.
66 However, according to the American Society of Appraisers (ASA), the income approach is often employed to assess the value of a collection of assets or individual machinery units that collectively function to produce a marketable product, thereby generating an income stream. This assemblage of assets, which includes both intangible and working capital assets, is commonly referred to as a business enterprise or going concern. A business enterprise encompasses all tangible assets, such as property, plant, equipment, and working capital, as well as intangible assets associated with an ongoing business. If the appraisal's purpose necessitates the exclusion of intangible value, the appraiser may need to evaluate the intangibles separately, particularly if they significantly contribute to the overall value (Cf. ASA [2011], p. 135).
67 Cf. White et al. [2003], p. 155.
68 RICS [2007], p. 2.
69 Corgel et al. [2001], p. 308.
70 Cf. White et al. [2003], p. 155; Cf. Metzger [2018], p. 121.
71 Cf. White et al. [2003], p. 156.
72 Cf. Petersen [2008], p. 46.

73 Cf. Sommer and Kröll [2017], p. 244ff; Cf. Leopoldsberger et al. [2016], p. 455.
74 Also see IFRS 3: Fair Value Measurements, para. 71, Fair Value Hierarchy and para. 86 Level 3 Inputs.
75 IfS [1999], p. 14.
76 Other than RCN, reproduction cost is the current cost of reproducing a new replica of the asset using the same, or closely similar, materials.
77 Cf. BVS [2024], p. 3.
78 Cf. SA [2011], p. 32.
79 Please also refer to TEGOVA [2022], p. 71. para 7.7 and ASA [2011], p. 34.
80 IfS [1999], p. 15.
81 Cf. BVS [2024], p. 3.
82 IfS [1999], p. 15.
83 Plock et al. [2025], p. 50.
84 Cf. ASA [2011], p. 38.
85 Cf. Ulaganathan and Sugumar [2024], p. 1
86 Cf. BVS [2024], p. 4.
87 Cf. Ellsworth [2009], p. 31.
88 Cf. BVS [2024], p. 4.
89 Cf. BVS [2024], p. 4.
90 Cf. ASA [2021], p. 45.
91 Cf. ASA [2021], p. 45.
92 Cf. ASA [2021], p. 45.
93 Cf. BVS [2024], p. 4.
94 Since trending is the most frequently applied method in the valuation practice, in this chapter the application of the trending method will be described in more detail than the detailed and the cost-to capacity method.
95 ASA [2011], p. 537.
96 ASA [2011], p. 545.
97 Cf. BVS [2024], p. 4.
98 Cf. BVS [2024], p. 4.
99 Cf. BVS [2024], p. 5.
100 Cf. IFRS 16, para. 16-22.
101 Cf. BVS [2024], p. 7; IfS [1999], p. 13.
102 Cf. BVS [2024], p. 7.
103 Also see ASA [2011], p. 56.
104 Plock [2025], p. 43.
105 Cf. BVS [2024], p. 8.
106 ASA [2011], p. 57.
107 Also see TEGOVA [2022], p. 135.
108 Plock [2025], p. 44.
109 ASA [2011], p. 56.
110 Cf. BVS [2024], p. 8.
111 Plock [2025], p. 45.
112 Also see TEGOVA [2022], p. 77.
113 Plock [2025], p. 46.
114 Cf. Metzger Bernhard [2018], p. 176.
115 Cf. TEGOVA [2022], p. 122; Also see: Renner and Sohni [2012], p. 197.
116 Cf. ASA [2011], p. 390.
117 Cf. ASA [2011], p. 517.
118 Plock [2025], p. 63.
119 IfS [1999], p. 18.
120 Cf. TEGOVA [2022], p. 184.

121 Cf. BVS [2024], p. 8.
122 Cf. BVS [2024], p. 8.
123 Also see TEGOVA [2022], p. 75.
124 Cf. BVS [2024], p. 8.
125 Cf. BVS [2024], p. 8.
126 Cf. TEGOVA [2022], p. 75.
127 Cf. Junius and Piazolo [2008], pp. 339ff.
128 Cf. TEGOVA [2022], p. 74.
129 Cf. TEGOVA [2022], p. 73.
130 Cf. BVS [2024], p. 4.
131 Cf. BVS [2024], p. 9.
132 Cf. Crane [2018], p. 3.
133 Cf. Crane [2018], p. 12.

7 Reporting, Documentation and Quality Check

The concluding phase of the valuation process, Phase V of the project, entails the preparation and documentation of the valuation report (see Figure 7.1). This report signifies the culmination of the valuers' endeavors and serves as the deliverable presented to the client upon the fulfillment of the assignment. Typically, the valuation model itself is not disclosed to the client, nor do they witness the extensive research analysis and calculations involved. Consequently, the report, along with any presentation that may be conducted, represents the most tangible output received by the client from the valuer.

Despite advancements in digital valuation techniques and the evolving nature of valuers' reporting methods, the preparation of a report remains a crucial component of the valuer's responsibilities. This report serves as the product or deliverable provided to the client, underscoring its significance in the valuation process.

Hence, in the following chapter, both the essential contents of a valuation report and the structure of its compilation will be described. Additionally, guidance on professional structuring and writing style will be provided, along with a description of the review and quality assurance process.

7.1 Reporting

Generally, the valuation report aims to present a comprehensive overview of the valuation and generally includes a description of the asset being valued and its relevant market, the methodology employed, all underlying relevant assumptions and, ultimately, the derivation of the valuation results. However, the report to be prepared by a valuer can vary greatly in terms of scope, structure and content depending on the valuer's engagement and role, which leads to different types of valuation reports. While external valuation purposes typically require a detailed full scope valuation report, internal or economical decision driven purposes may allow for shorter, more informal reporting. In the latter context, elements of valuation consulting may play a role, providing the client with strategic assistance through the valuation report, for instance, in making investment decisions.

However, this is always subject to individual case decisions and upon agreement with the client and potentially other stakeholders involved.

7.1.1 Type of Valuation Report

As shown in Figure 7.2, valuation reports in Fixed Assets valuation can be categorized into three primary types, each tailored to specific assignments and purposes: Valuation Report,

DOI: 10.1201/9781003622413-7

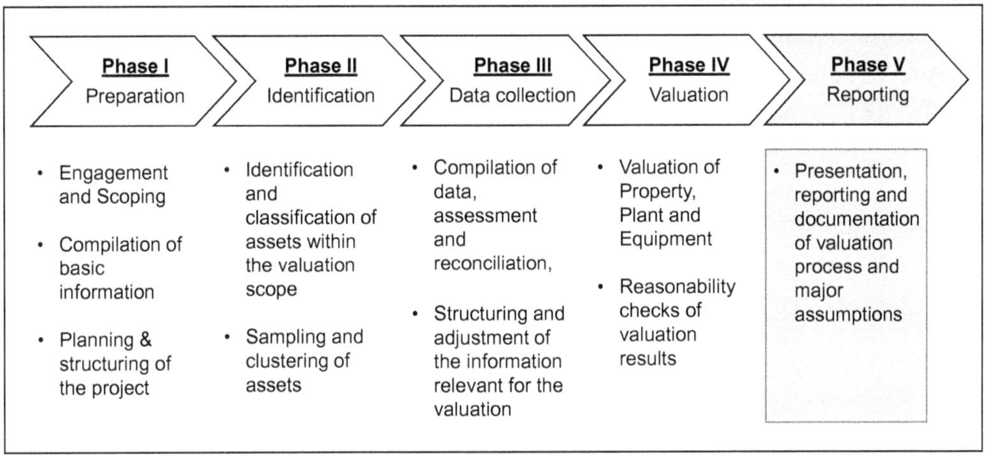

Figure 7.1 Five Phases Approach to Fixed Asset Valuation.
Source: Adopted from Plock et al. [2025], p. 20

Valuation Review and Valuation Consulting. The format and content of these reports are influenced by the role of the valuer and the intended use of the valuation findings.

7.1.1.1 Valuation Report

A Valuation Report, often referred to as a Full Valuation Report, is the most comprehensive form of reporting. It provides an exhaustive analysis of the asset being valued, incorporating detailed descriptions, relevant market analyses and methodological explanations. This report is typically structured in an academic style, integrating references to literature and legal precedents to substantiate the valuation both technically and legally. The full scope of the report ensures a thorough examination of the asset, making it suitable for assignments requiring detailed insights.

The subsequent discourse primarily pertains to the preparation of Full Valuation Reports wherein the valuer acts in the capacity of an independent expert. This scenario may be considered the standard case, and therefore, the standard report format is delineated herein. Any deviations from this approach must align with the defined scope of work and be duly coordinated with the client. A broad distinction and related requirements can be made between internal and external valuation purposes.

7.1.1.2 Valuation Review Report

Valuation Review Reports are often presented in a somewhat less formal style, resembling a memorandum format. These reports focus primarily on the methodology and fundamental assumptions used in existing valuations, leading to the derived value. The aim is to verify the accuracy and reliability of the valuation assumptions and methodologies. Valuation Review Reports are critical for ensuring that the valuation process adheres to accepted standards and practices.

Area of Valuer's activity	Related report and documentation
Full Valuation	• Comprehensive formal valuation report • Type and scope of the report depends on the reason for the valuation and its related specific requirements and intended usage • Full scope vs. limited scope might significantly impact the level of detail of the report
Valuation Review	• Typically, documentation is in a less formal report style and more like a Valuation Memo • Depends on scope of work and use of the report
Valuation Consulting	• Variety of consulting related reports • Typically, less formal than a comprehensive valuation report • Depends on scope of work and use of the report

Figure 7.2 Area of Valuer's Activity and Types of Valuation Report.
Source: Authors.

7.1.1.3 Valuation Consulting Report

Valuation Consulting Reports are less formal and resemble consulting documents. These reports emphasize strategic recommendations based on the professional view of the valuer rather than the objective assessment of valuation objects. In consulting assignments, the valuer often acts as an advisor, providing clients with clear and strategic guidance tailored to their specific needs. Hence, the language and style of these reports reflect the advisory role, offering actionable recommendations that support the client's economic decision-making processes.

Each type of report described above serves distinct functions and is adapted to meet the specific requirements of the assignment, ensuring that clients receive the most relevant and actionable insights based on the purpose of the valuation.

7.1.2 Report Writing Process

The entire valuation and report writing process, and particularly the possible degree of its standardization, depends heavily on the scope of the valuer's responsibilities and type of work performed. In valuation firms where a large number of recurring valuations of similar standard properties are carried out, the degree of standardization is very high and will increasingly involve machine-based and digitalization processes. In most common standard valuation programs, reporting nowadays is already an integral part of the software and thus the valuer is less involved in actually writing the report but tends to take

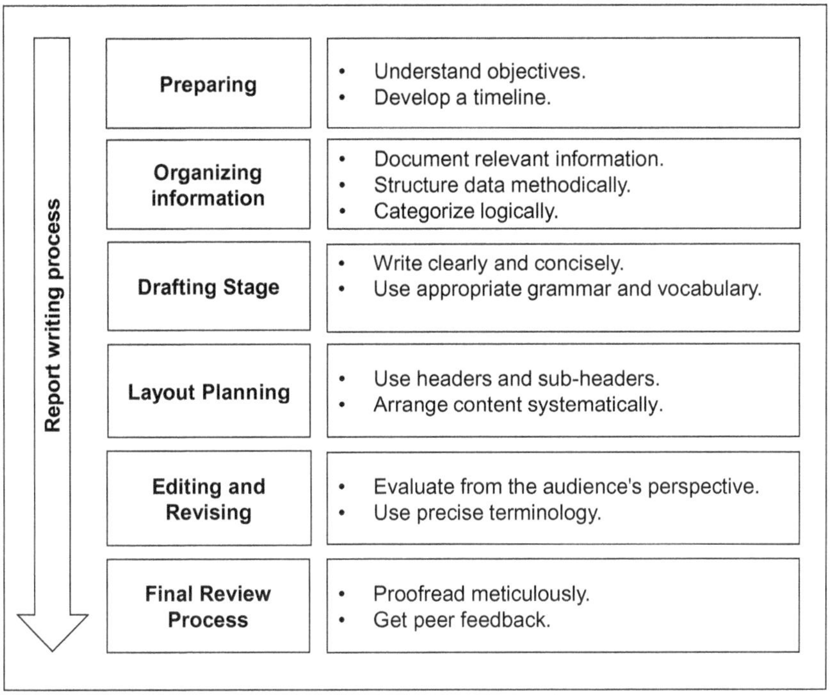

Figure 7.3 Report-Writing Process.
Source: Authors.

on the role of assessor and reviewer rather than having to create and structure a report "from scratch."

The situation may be different for valuers who are more frequently involved in the valuation of special properties or whose valuations are carried out in the context of more complex and, above all, individual valuation issues. For such special projects, which do not recur regularly, individual text modules can be reused, but the reports often need to be restructured and rewritten.

As shown in Figure 7.3, the preparation of a valuation report is a meticulous process that begins with comprehending the underlying objectives of the report itself. Understanding these objectives is crucial as they shape the direction and focus of the entire document. Once the objectives are clear, it is essential to develop a comprehensive timeline that will guide the progression of the report. This timeline serves as a roadmap, ensuring that all necessary steps are completed in a systematic and timely manner.

Following the initial preparation of the report writing process, the next step involves gathering and organizing information. It is imperative to accurately document all pertinent information available, as this forms the foundation upon which the valuation is based. The collected data must then be methodically structured, allowing for easy access and reference throughout the drafting process. Categorizing the information in a logical manner facilitates coherent analysis, enabling the valuer to draw meaningful conclusions that will be conveyed in the report.

The drafting stage focuses on composing the document with clarity, simplicity and brevity. The aim is to communicate complex valuation analyses in a manner that is easily understandable to the intended audience. Employing appropriate grammatical structures and vocabulary suitable for the academic context of a professional valuation report is necessary to maintain the credibility and professionalism of the report.

Once the draft is complete, attention turns to layout planning. Organizing the report into distinct sections, utilizing headers and sub-headers, helps delineate key themes and enhances the reader's ability to navigate the document. Systematically arranging paragraphs, tables and diagrams is equally important, as it improves readability and comprehension, allowing the audience to easily grasp the presented information.

Editing and revising are critical stages that involve assessing the draft from the perspective of the intended audience. Utilizing precise terminology and maintaining an appropriate tone throughout the document reinforces the report's authority and reliability. Especially if the report is written not only by one valuer but an entire project team which is involved in the drafting of a report, a uniform editing process is key.

The final review process is the concluding step, where meticulous proofreading of the final draft is conducted to catch any errors or inconsistencies. It is highly beneficial to solicit feedback from a colleague or peer, as their insights can help ensure accuracy and clarity. This collaborative approach enhances the quality of the report, ultimately delivering a polished document that effectively serves its intended purpose.

7.1.3 Content of Report

Many national and international valuation standards and guidelines contain clear requirements for the content of valuation reports. These requirements are part of the standards and must be complied with when a valuer is tasked with conducting a valuation in accordance with these guidelines. The RICS Global Valuation Standards specify clear requirements for the structure and content of a Fixed Assets valuation report. Consequently, it is crucial for the valuer to ensure that their report aligns with the requirements of the valuation engagement and that the content is adequate for its intended purpose.[1]

According to RICS, the valuation reports must address the following matters[2]:

- Identification and status of the responsible valuer
- Identification of the client and any other intended users
- Purpose of the valuation
- Identification of the asset(s) or liability(ies) valued
- Basis(es) of value adopted
- Valuation date
- Extent of investigation
- Nature and source(s) of the information relied upon including sources of key data and inputs used
- Assumptions and special assumptions
- Restrictions on use, distribution and publication of the report
- Valuation approach and reasoning, including any valuation method(s) and complex or proprietary model(s) used
- Amount of the valuation or valuations
- Date of the valuation report

- Commentary on any material valuation uncertainty (MVU) in relation to the valuation where it is essential to ensure clarity on the part of the valuation user
- A statement setting out any limitations on liability that have been agreed
- Significant environmental, social and governance (ESG) factors used and considered

Most valuation reports, particularly full scope valuation reports, primarily adhere to the contents outlined by RICS and IVS. In the case of limited scope analyses, certain elements may be omitted or explained less extensively, which must be justified separately. For valuation review memos, some content may differ, but most will be similar. Consulting reports may significantly deviate in content, as they are not valuation reports but rather advisory documents.

Besides the content of the report, the structure is also critical for ensuring clarity and coherence, as it enables readers to navigate the document with ease and locate necessary information efficiently.

7.1.4 Structure of the Report

Selecting an appropriate logical framework for the valuation report and its intended audience is essential. Various reports and audiences may require different structural approaches, such as chronological, comparative analysis or problem-solution formats. A carefully chosen structure facilitates the reader's journey through the valuation content, ensuring that the narrative or argument develops clearly and engagingly.

As depicted in Figure 7.4, most valuation reports are structured into three distinct sections: the Executive Summary, the Main Report and the Appendix. The Executive Summary is specifically crafted for top executives and contains key "elevator points" that succinctly convey the essential findings of the report. The Main Report is intended for management or principal readers and presents a comprehensive opinion and valuation. Lastly, the Appendix is designed for analyst readers and supplies additional detailed information that supports the opinion and valuation.

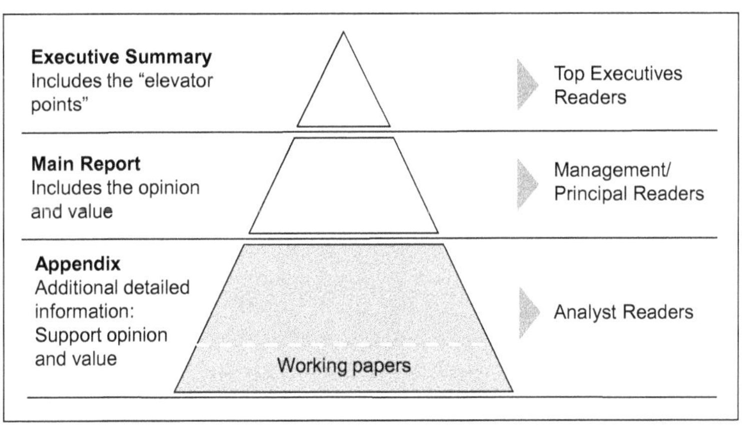

Figure 7.4 Valuation Report Structure.
Source: Authors.

The valuation report itself generally follows a consistent structure. The Royal Institution of Chartered Surveyors (RICS), in alignment with the International Valuation Standards (IVS), offers a standardized structure that serves as a valuable guideline. If the report is prepared in accordance with the RICS Red Book, the valuer must follow this structure. Typically, the report will be organized into major sections, although not necessarily in this order[3]:

a Introduction
b Purpose and Basis of Value
c Assumptions and Special Assumptions
d Subject of Valuation
e Description and History of the Asset(s), and the Business Entity in which they have been utilized
f Accounting and Accounting Policies
g Financial Statement Analysis, if applicable
h Business and Marketing Plan Analysis, and Prospects
i Search Results for Comparative Transactions
j Industry Economic Context and Environment, Yields and Risk Assessment
k Valuation Methods and Conclusion
l Caveats, Disclaimers and Limitations

Most valuation reports, particularly full scope valuation reports, are predominantly aligned with the structures outlined by RICS and IVS.

7.1.5 Report Writing

As described above, the style of most valuation reports tends to be academic, aiming to describe the facts accurately and explain the valuation approach used in accordance with the relevant valuation standard.

Even if a valuation report is preceded by an executive summary containing the most important information about the subject of the valuation, the valuation method and the valuation result, as shown above, the structure of the valuation report is more academic than that of the consulting report, as shown in Figure 7.5.

The individual assumptions are explained, and the value is derived from them. In a valuation consulting report, both the structure and the writing style are slightly different. The structure is more results-oriented and the writing style is more in the form of recommendations for action and advice. The following points may be helpful in preparing valuation reports, memos and consultancy reports.

Logical Grouping

Organizing related valuation information within the same section of a report is crucial for coherence and clarity. Before drafting a valuation report, it's helpful to thoughtfully plan how to cluster all pertinent points, such as asset characteristics, valuation methods and assumptions. This strategic grouping ensures that each section flows logically and that connected ideas are easily accessible, thereby enhancing the reader's comprehension and retention.

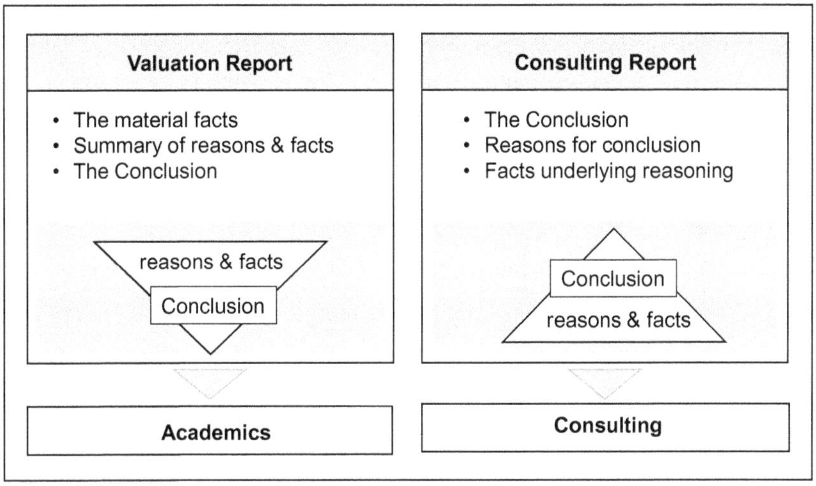

Figure 7.5 Academic vs. Consulting Report Structure.
Source: Authors.

Prioritizing Key Points

An effective report presents its purpose, main valuation conclusions, methodologies or recommendations at the outset. This approach immediately informs the reader of the report's intent, enabling them to grasp the core findings without delay. Deviating from this order should only occur if there is a compelling rationale, such as needing to establish context before arriving at a conclusion.

Choosing an Appropriate Structure

Selecting the right logical framework for the valuation report and audience is vital. Different reports and audiences may necessitate varied structures—such as chronological, comparative analysis or problem-solution formats. A well-chosen structure guides the reader smoothly through the valuation material, ensuring that the narrative or argument unfolds in a clear and engaging manner.

Applying the Magic 7 Rule

Cognitive psychology suggests that individuals can remember approximately 7 (+/-2) items at a time. To enhance readability and retention, limit sections, bullet point and headings to no more than seven items. If this number is exceeded the content should be divided into smaller, more manageable groups. This approach aids in organizing complex valuation information and supports the reader's cognitive processing.[4]

Engaging in a Question-and-Answer Dialogue

A well-crafted valuation report resembles a conversation between the reader and the text. As readers progress through the material, they naturally generate questions about the valuation process and conclusions. A responsive text anticipates these inquiries and provides answers, maintaining the reader's interest and ensuring that their informational needs are met as they read.

By embracing these principles, valuers can produce reports that are not only informative but also engaging and easy to understand, ultimately improving communication effectiveness within professional settings. Through careful organization, clear expression and thoughtful engagement with the reader's needs, valuers can achieve mastery in business writing tailored to fixed asset valuation.

7.1.6 Transparency in Reporting

One of the most important features of making a good and comprehensive valuation report is a transparent and understandable presentation of how the key valuation parameters and assumptions have been derived. However, this part of the report is often neglected. In practice, a great deal of space is sometimes devoted to the description of the property and general descriptions of the local markets, with only a brief presentation of the assumptions and the derivation of the value at the end of the report. Although the description of the property and a well-founded positioning within the relevant market are central and important components of the report (as outlined above) in order to substantiate and justify the derived value, a transparent outline of the valuation assumptions is essential. In particular, as the valuation parameters can never be considered in isolation but only in combination, a transparent description of the applied parameter is important to understand if those parameters are reasonable and appropriate.

7.2 Quality Assurance

A fundamental aspect of report preparation is quality assurance. The internal quality check of a valuation report concerns both the valuation itself and the formal structure of the evaluation report. Two elements are therefore examined within the framework of quality assurance: whether the valuation is accurate and meets the formal criteria relevant to the valuation context, and whether the valuation report is correct in terms of its formal and linguistic presentation.

7.2.1 Quality Assurance Process

The content of the valuation is reviewed basically in a similar way to an external valuation review as described in Chapter 7, but usually less formal and detailed. As the quality assurance reviewers are typically the project managers plus a member of the management team, they are involved in the valuation process from the outset, particularly in complex projects, and therefore do not usually start from scratch as in an external review. The focus of the internal content review is therefore usually on the key valuation assumptions, the plausibility of the valuation result and the formal requirements for the report. In addition, the report is reviewed for content, form and linguistic.

The time required for quality assurance within the entire report writing process should not be underestimated. The work that remains to be done after the valuer has finished a first draft can be considerable. Depending on the type of project (standard assessment vs. special project), several "review loops" may be necessary, as shown in Figure 7.6, with several repeated reviews of the various quality assurance measures. This has to be taken into account in the project planning and the deadline agreed with the client.

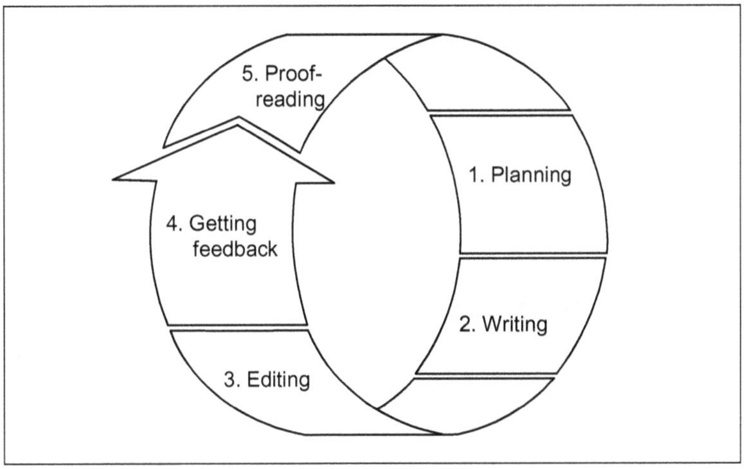

Figure 7.6 Report Quality Assurance Process.
Source: Authors.

7.2.2 Draft Report Version

The report is usually first drafted and sent to the client for approval. This is usually done so that the client has the opportunity to read the report at the draft stage and check, for example, that all the property information they have provided, such as floor areas or cubic volumes, is correct and complete and that there are no misunderstandings between the client and the valuer. It is important that this draft version is clearly marked as such. Usually each page will have a "Draft" watermark and a disclaimer on the title page. A disclaimer needs to clearly highlight the preliminary draft status of the report and emphasize that the report is therefore subject to further changes and adjustments. A draft report should therefore not be used as a basis for decision making by the client but should be taken with appropriate caveats. A typical disclaimer for a draft report might be:

> "Preliminary draft report, subject to further possible review, work and revision. Should not be relied upon for any purpose whatsoever."

After the client has read the draft report, there is usually a brief consultation between the client and the expert to answer any questions or comments the client may have and to clarify any outstanding issues. Once all issues have been resolved, the valuer will produce the final version of the report.

7.2.3 Final Report

Once the draft valuation report has been thoroughly reviewed and all feedback from the client has been incorporated, the final valuation report is prepared. This version of the report represents the completed, reviewed and approved document that will serve as the official record of the valuation.

The final report is formatted professionally, without any watermark or disclaimer that characterized the draft version. It is essential to remove any indications of preliminary status, as this report is intended to be relied upon for decision-making purposes. The final valuation report should include any relevant certifications or endorsements, such as the signature of the valuer and, if applicable, a professional seal or stamp to verify its authenticity and completeness.

Upon completion, the final report is formally delivered to the client, often accompanied by a discussion or presentation to ensure the client fully understands the content and implications of the report. The delivery of the final report marks the conclusion of the valuation assignment, providing the client with a definitive resource for making informed decisions regarding the property in question.

Notes

1 E.g. see. Australian Government [2022], p. 9.
2 RICS [2024], p. 68; also refer to IVS 106 Documentation and Reporting, p. 60 in RICS [2024] Global Standards.
3 RICS [2024], p. 112.
4 Mille [1956].

8 Third-Party Valuation Review

As already described in Chapter 2, the work of the valuer does not only consist of preparing their own valuations, but also of reviewing and assessing valuation reports prepared by third parties. This process is referred to as a valuation review.[1]

When conducting valuations, various risks and sources of error can arise, which must be identified as part of a valuation review. These may include formal errors in the selection of the valuer who may not have the required qualification for the specific valuation task, or the type of report, as well as methodological errors in the choice of approach and valuation procedure, and implausible assumptions regarding the input parameters of the valuation. Particularly with the latter, there is often no clear right or wrong; instead, assumptions are frequently subject to the appraiser's market assessment and must be scrutinized for their plausibility.[2]

This chapter explains the most common sources of error (regardless of whether they were caused deliberately or unintentionally) in the preparation of fixed asset valuations. This chapter also includes a structured checklist for conducting valuation reviews.

8.1 Focus Areas of Fixed Assets Valuation Review

8.1.1 Formal Requirements of Valuer

The valuer must fulfil the formal requirements that are necessary for the respective valuation to be carried out.[3] These are primarily based on the reason for the valuation or the valuation background. In practice, the formal requirements usually relate less to an economic valuation reason but are more important when carrying out statutory valuations which serve a regulatory purpose.

In many countries e.g. valuations for tax or regulatory purposes may only be carried out by state-certified valuers.[4] Banks also often have formal requirements regarding the status of the valuer for valuations carried out for financing or lending purposes.[5] The same applies to valuers who wish to prepare valuation reports for financial reporting purposes. In individual cases, it must be checked whether both the valuer and the valuation company fulfil the respective formal requirements.

8.1.2 Unclear Definition of the Determined Value

The first consideration in reviewing the quality of a valuation report applies to the type of derived value. In most cases, in a transaction or for tax purposes the Fair Value or Market Value is to be determined. As explained above, different values can be determined for one

DOI: 10.1201/9781003622413-8

valuation object, which sometimes differ significantly from each other because of their definitional assumptions. A general rule is that the investment value of an asset is more likely to be higher than the Fair Value and the Market Value, and both in turn are higher than the Liquidation Value.[6]

The investment value often includes subjective moments, which reflects the individual buying interest of the potential investor. This could be, for instance, expected synergies between the assets to be purchased and one's own company, but also targeted value appreciation measures to extraordinary investments or restructuring. These considerations play no role for the Fair Value and the Market Value, which are subject to the principle of continuing use or the principle of going concern. The Fair Value is derived by objective market conditions and the exclusion of subjective factors.

The liquidation value, on the other hand, requires an immediate sale under timely enforcement, which excludes a reasonable marketing period of the valuation object. Therefore, it can be assumed that the price that may be attained in the market will be lower. Even individual assets, such as a plant, are worth less than assuming their continued, consistent, operational usage (going concern).

Although the value is usually defined within the report, it is not always evident whether it actually determines the definition requirements. For example, the derived Fair Value is sometimes based on a business plan, which already considers anticipated future synergies and reflects them monetarily in the cash flow. This would eventually lead to an overestimation and a Fair Value that is too high. Therefore, it must be ensured that the appraiser always selects the correct value for the valuation purpose,[7] defines it in the report and derives it according to the definitional assumptions.

8.1.3 Valuation Scope and Valuation Object

In most cases, valuations are made to determine the Market Value of the company (equity) or individual assets. Essential is the distinction of whether in the context of valuation the value of all assets (total assets) of an enterprise or the equity of all assets, the company value, is to be determined.

During the valuation, the valuation object should be chosen correctly, and it should be decided whether only individual assets or the entire company should be assessed. It should be noted that the total of assets does not have to equal the value of the whole company, as the goodwill must be regularly taken into account. It is also important that all intangible assets are identified and included in the valuation.

In the process of a target's valuation for a transaction or the forming of a Joint Venture it should be ensured that the underlying business plan and the related forecasted cash flows are determined on a stand-alone basis. Future individual synergies can and indeed should be considered in pricing but must not be mixed with an objectified Market Value of the target. Before considering whether or not the valuation subject has been "properly" valued, one should consider whether the "right" valuation subject has been valued.

8.1.4 Applied Valuation Methods

As stated above, the correct selection of the applicable valuation method depends on the one hand, on the valuation purposes and on the other hand, on the valuation object. Theoretically, valuation is supposed to lead to the same result using any method, but practically this is rarely true. For certain valuation objects, and for certain valuation purposes, certain assessment methods need to be applied.[8]

Thus, the choice of the method can already have, in practice, an impact on the value to be derived. In a difficult market environment, the resulting value by means of the replacement cost method can be higher because of the lack of direct reference to the current market situation.[9] Another issue, which matters in the same context, is market fluctuations. Below, some of the limitations of the method observed in practice will be discussed.

8.1.4.1 Application of the Market Approach

Although the market transaction method is the simplest and most direct method, it is considered in the literature to be the best and preferred method because of its immediate market dimension.

This advantage also forms the weak point of the method because there must be a sufficient number of market-comparable transactions or multiples available in order to apply it. As outlined in Figure 8.1, those applied transaction comparables are also the major area of potential pitfalls in the market approach.

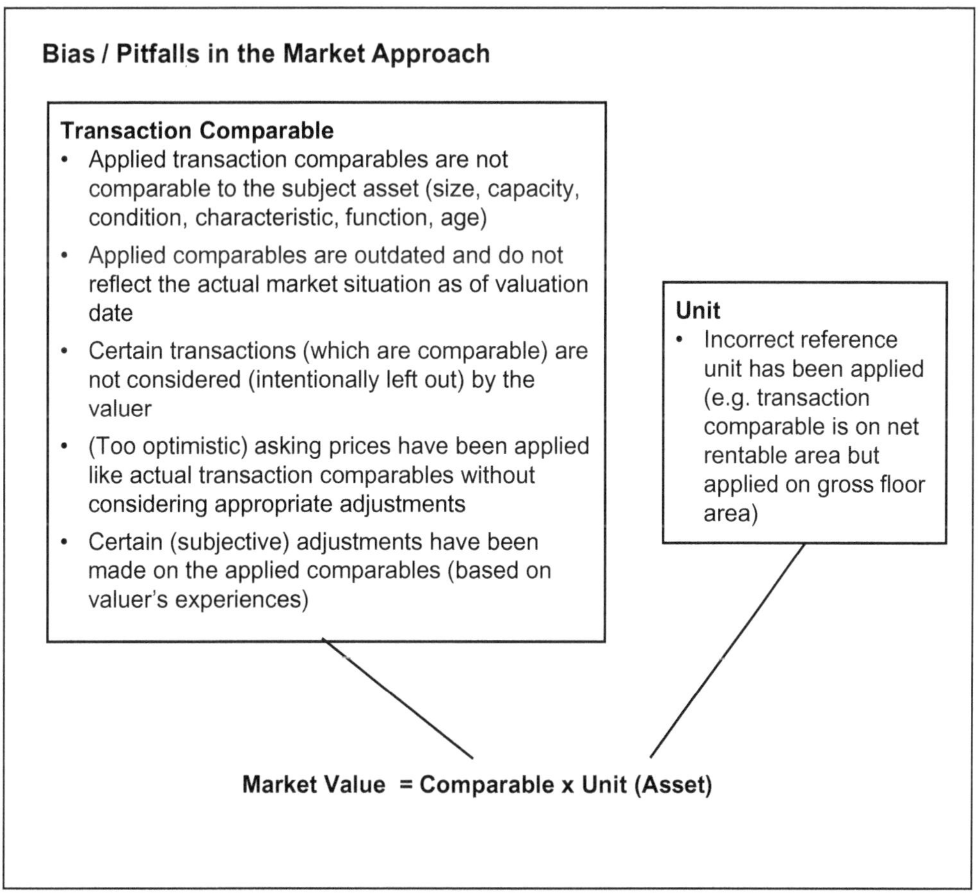

Figure 8.1 Common Pitfalls in Applying the Market Approach.
Source: Authors.
Note: Adapted from Damodaran [2025].

Extensive market research that considers project-specific factors seems therefore essential for the value derivation via the comparative method.

8.1.4.2 Application of the Income Approach

The income approach is a robust method for market-based valuation. It primarily applies where no direct comparison transactions are available, or due to the reason for the

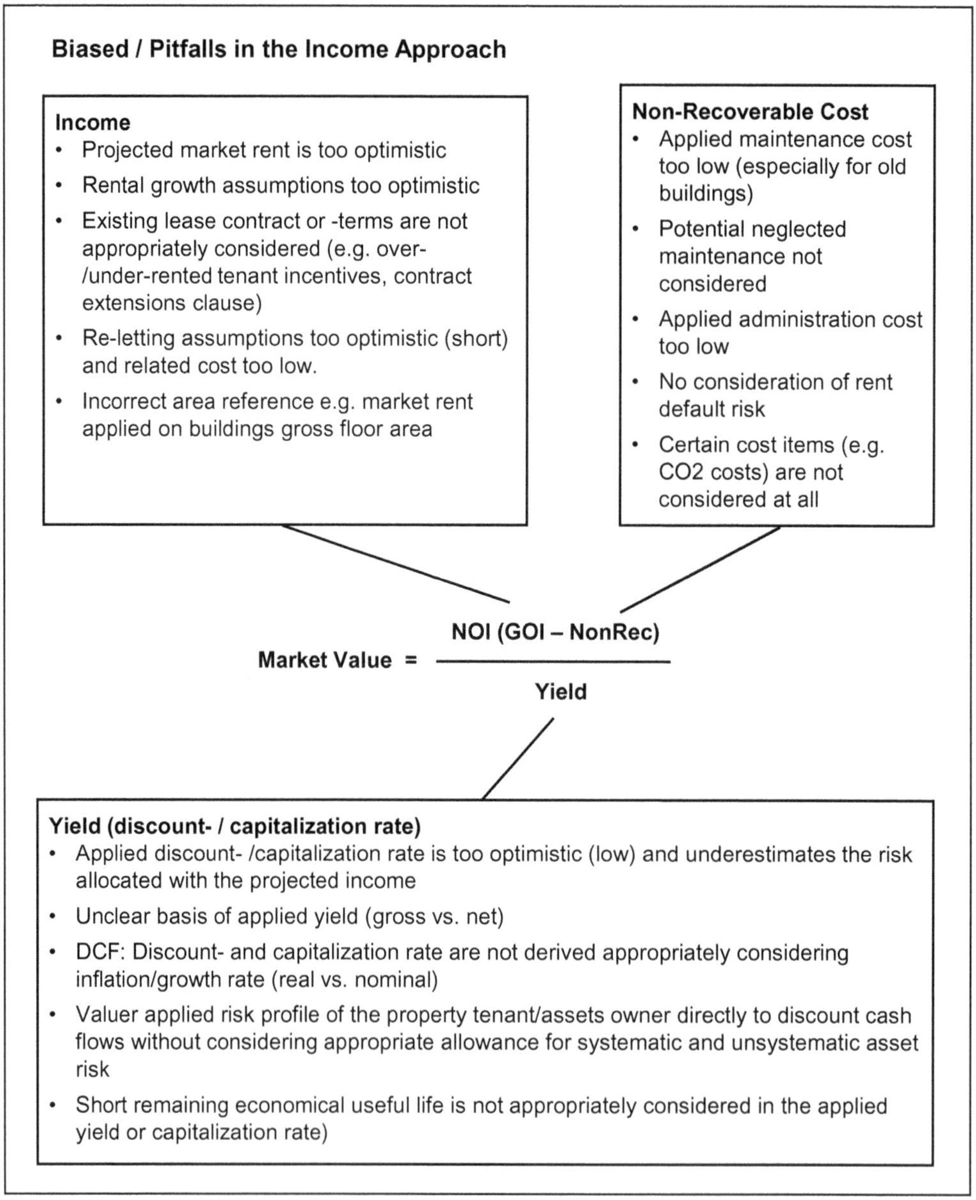

Figure 8.2 Common Pitfalls in Applying the Income Approach.
Source: Authors.
Note: Adapted from Damodaran [2025].

valuation, existing budget figures must be considered. The great detail displayed in the determinants of the DCF method allows an explicit illustration of all influencing factors.

When reviewing a valuation based on an income approach, despite the utilized method, when assessing the assumed parameters a particular focus should be on the projected future income and the applied yield. Especially the yield should be transparent derived and reflecting the risk associated with the projected income, since already small changes in the yield have a major impact on the value (Figure 8.2).[10]

8.1.4.3 Application of the Cost Approach

The substance-oriented approach is especially applied in valuation practice of Fixed Assets. The cost approach can be a robust method especially for the valuation of Machinery and Equipment asset. However, it has to be considered that the Market Value is not based on what an object would cost today, but rather what the market is willing to pay for at the valuation date. Especially in the valuation of Fixed Assets with underutilization or technical and economic obsolescence, considering pure depreciation of the replacement cost values to derive the Fair Value is not appropriate. Obsolescence and low utilization must explicitly be taken into account, and it is recommended that the reasonability be cross-checked by undertaking a profitability test. Common mistakes and pitfalls of the Cost Approach are shown in Figure 8.3.

8.1.5 Subjectivity of Key Parameters

Key parameters are so-called value drivers, since small changes may have disproportionate effects on the valuation result.

For the market approach, it is important to set a sufficient justification of the comparison data (comparables), such as comparative transactions or multiples. Especially when adjustments to the comparables are made in deriving the value, this must be justified coherently.

The value drivers of the income approach primarily consist of the assumed revenue growth, margins, the discount rate and the so-called terminal growth rate, which is the assumed growth at the end of the planning period. The plausibility of these assumptions should be validated with historical data and supported by appropriate market analyses.

In the course of the cost approach, the chosen replacement values and the remaining useful lives are the main value drivers, whereby the approaches for obsolescence should also closely be paid attention to. For all valuation methods, particular caution is advised if the valuation expert justifies significant assumptions with his or her expert's experience. While for many minor assumptions, experience may be justifiable, for key value drivers, reliable market data must be obtained and, if necessary, one's own research should also be carried out.

8.1.6 Transparency in the Preparation of Valuations

Often challenge in the valuation of assets results from the limited availability of reliable market information. This refers to all three described valuation approaches, and with regard to the performed valuation often results in subjective estimates and assumptions made by the appraiser.

Due to this specific market transparency challenge, a coherent description of the procedure of the entire valuation process is particularly necessary, and a detailed explanation

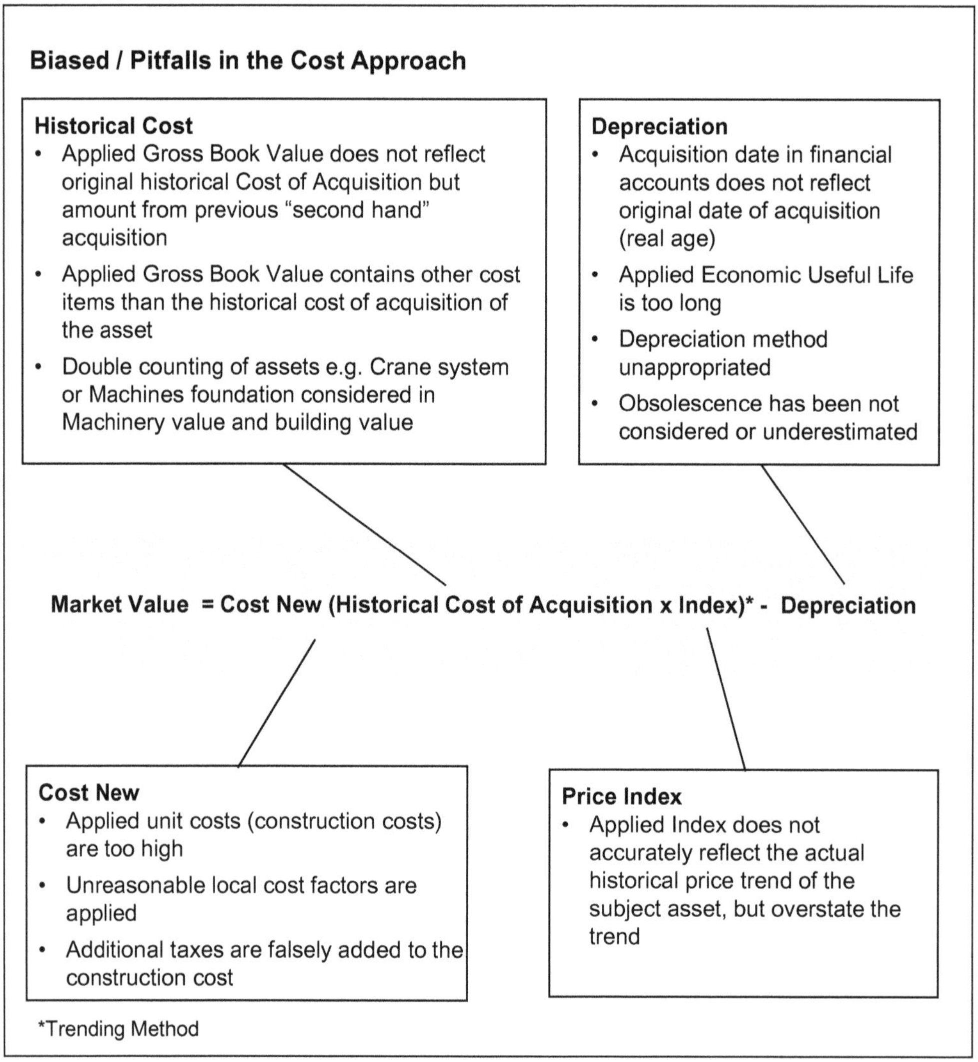

Figure 8.3 Common Pitfalls in Applying the Cost Approach.
Source: Authors.
Note: Adapted from Damodaran [2025].

of all value-influencing assumptions are recommended to get plausible valuation results. To ensure adequate transparency of the report, it is advisable to pay attention to the following contents as part of a valuation report[11]:

a Identification and status[12] of the valuation expert
b Identification of the client and of other addressees
c Purpose and background of the valuation
d Identification of the valuation subject (asset or liability) to be valued
e Basis of value
f Valuation date

g Scope of the investigations
h Type and source of the used information
i General and special assumptions
j Restrictions on the use, distribution or publication of the report
k Confirmation that the engagement for the valuation has been executed in accordance with the relevant valuation standards (e.g. IFRS, IVS, RICS)
l Description and support of utilized valuation approaches
m Clear and complete presentation of results
n Date of the valuation report and signature of the valuation expert
o Other restrictions

In addition to this formal structure, a coherent justification of the chosen valuation method and the documentation of the derivative input parameters based on appropriate market data are equally important.

8.1.7 Quantity Structure and Key Assumptions

The basis of each valuation is the underlying quantity structure. Depending on the valuation object, this can primarily be balance sheets or financial forecasts, contractual constructs, land and leases (property valuation) or technical specifications (Machinery and Equipment).[13]

It must be ensured that the quality of financial information of local companies does correspond to international standards and always reflects the actual state of the company.[14] If obvious doubts about the plausibility of the figures arise, appropriate tests should be carried out. To exclude such doubts and to minimize risks in a transaction, extensive analysis, so-called financial and legal due diligence reports, is carried out by the potential buyer and his advisors. These reports often provide a solid first overview of the financial situation and also of the quality of the available data.[15] If the data is not robust enough to carry out a valuation and if appropriate, further tests must be carried out to define a sufficiently solid quantity framework.[16]

8.2 Plausibility Assessment of Valuation

To assess the plausibility of appraisal reports and verify the Fixed Asset value derived within them, depending on the valuation context, the object being valued and the market environment three different approaches are commonly employed in practice.[17] This three approaches include: A review of the third-party valuation results based on market indicators, reviewing the methodology and ensuring the plausibility of the quantity structure and key assumptions or performing an independent own comparative valuation.[18]

8.2.1 Review of the Valuation Results Based on Market Indicators (Output)

A relatively straightforward and efficient method for validating a property's value involves using comparable transactions that occurred near the valuation date.

This approach is generally sufficiently precise when dealing with market-standard valuation objects for which there is an active market. However, for complex valuation objects or properties whose value is derived from a complicated income structure, this method is often inadequate.

Additionally, due to the limited ability to adequately account for property-specific characteristics, this method is typically applied in practice as a rough approach for approximate verification. It serves as an initial check rather than a definitive assessment, providing a general sense of whether the initial valuation aligns with observable market trends.[19]

8.2.2 Reviewing the Methodology and Ensuring the Plausibility of the Quantity Structure and Key Assumptions (Input)

The basic idea behind this approach is that, if the correct valuation method is chosen, all relevant input parameters are market-appropriate and the calculations are accurate, the resulting valuation should be correct.

The reviewer must critically engage with the valuation parameters and assess whether the values fall within a reasonable range. The assumptions and judgments made during the valuation process should undergo a thorough analysis. Since the reason for the property valuation is the risk of a potential default, valuation parameters should be determined with caution, ensuring that the aspect of security realization is adequately considered. Utilizing discretion solely to manipulate the parameters is not appropriate.

Due to the detailed analysis of each valuation parameter, this method of validating a valuation report can be significantly more complex and labor-intensive than the previously described approach.[20]

Reviewing calculations in a complex valuation model can be very time-consuming. Furthermore, there is often a practical issue where appraisers do not provide their detailed calculations. Consequently, in the validation process, recreating the entire valuation model is often necessary, which effectively equates to conducting an independent valuation.

8.2.3 Performing an Independent Own Comparative Valuation (Cross-Check)

To assess the plausibility of a valuation, conducting an independent valuation (optionally with the assistance of external experts), can be effective. If there are significant discrepancies between the two property values determined using the same method in the comparative valuation, the cause should be analyzed based on the valuation parameters applied.

Conducting a comparative valuation represents the most comprehensive method for plausibility checking. It is particularly recommended for complex valuation objects and whenever a sufficient number of comparable transactions is unavailable.[21]

8.3 Checklist for the Plausibility Assessment of a Fixed Asset Valuation

The following checklist can be used as a guide for a high-level plausibility check of valuation reports.[22]

---

1. Formal requirements and engagement

- a Is the valuation firm certified and authorized to carry out the valuation?
- b Is the valuer sufficiently qualified to carry out the valuation?
- c Is the report completely submitted including relevant working paper and notes?
- d Is the engagement letter attached?
- e Is the expert opinion signed (and sealed)?

f What value (pursuant to what definition) was determined, and is this appropriate for the valuation occasion?
g Has an on-site inspection of the valuation object, as well as management interviews, been conducted?
h What data (quantity structure) were available for the valuation and who provided this (sources)?
i Is financial information audited to the valuation date (if required for statutory valuation)?
j At what valuation date was the valuation conducted?
k Has the valuation been performed in compliance with the relevant (valuation) standards? e.g. IFRS 3/IAS 16/IAS 40, White Book (IVS), Blue Book (TEGOVA), Red Book (RICS), other relevant local standards.
l Are the valuation scope and valuation process adequately described? Are any items excluded from the scope of the valuation analysis, such as Fixed Assets at minor locations, work in progress, office equipment, etc.?
m Did the expert act as an independent valuer or as an advising consultant?
n Is there any documentation of access limitations issues (consider whether access limitations may materially impact value conclusions)?
o Does the valuation report state any assumptions (general or/and special) which are affecting / limiting the valuation result?

2. Valuation methods

p Which valuation approach was used (market, income or cost approach) to value the asset?
- Is the method applied appropriate for the valuation object (e.g. marketability) and the market environment (e.g. active market)?
- Is the method for the valuation purpose allowed?
- Has the method been applied correctly?

q Were multiple methodologies used? If so,
- How was the concluded indication of value arrived at?
- Is the spread in the various indications of value by methodology significant?
- If a single methodology is used, note why and whether or not the appraiser attempted to corroborate conclusions using an alternative methodology.

3. Significant key value drivers

The following key value drivers must be well reasoned and conclusive. If necessary, individual market research should be carried out for cross-comparison.

Market Approach

r Are the recognized comparative transactions close in time to the valuation date, particularly with regard to the prevailing market volatility?
s Are the applied comparative transactions sufficiently similar to the valuation object?
t Are the key figures of the comparable assets (multiples) sufficiently comparable with the assets being valued?
u Are adjustments transparently listed and conclusively substantiated?

Income Approach

- v Is the underlying business plan/rental projection plausible, particularly in terms of estimated growth figures and the profit margin?
- w Is the income projection considering/adequately reflecting any existing lease contracts and related terms?
- x Are the applied non recoverable costs (outgoings) reasonable and derived transparently?
- y How is the all-risk yield / discount rate / capitalization rate derived and is it reasonable?
- z Is the Terminal Value at the end of the detailed planning period plausible?
- aa For assets with a short remaining economic useful life, how has this been considered/reflected in the valuation?

Cost Approach

- bb What sources were used to determine the replacement cost new / reproduction cost new?
- cc Are any other value-related factors such as technical, economic or functional obsolescence taken into account?
- dd Is the scheduled remaining (economic) useful life of the asset appropriate?
- ee Are conducted modernization and repair measures or potential existing neglected maintenance taken into account?
- ff Were the chosen parameters logically set in relation to each other?

4. Arithmetic correctness

- gg Is the valuation model reasonable and properly structured?
- hh Is the value derived arithmetically correctly?
- ii Have random plausibility calculations been performed?

5. Report and documentation

- jj Have the outcomes of the valuation model been transferred correctly to the expert opinion (particularly in updated versions)?
- kk Do the parameters in the report correspond with those of the valuation model?
- ll Is the valuation report provided sufficient for the valuation purpose?
- mm Does the valuation result make sense (common sense)?

Notes

1 Cf. HKIS [2017], p. 31.
2 Also see: Hackelberg [2018], p. 362.
3 Cf. E.g. Australian Government [2022], pp. 5–7.
4 Cf. E.g. Australian Government [2022], pp. 5–7.
5 Cf. Australian Bankers Association [2019], pp. 2–3.
6 Hackelberg [2018], p. 362.
7 Cf. E.g. Chinese Valuation Standards [2013], Article 19.
8 Hackelberg [2018], p. 364.
9 Australian Government [2022], p. 10.
10 Cf. Gelter and De Neufville [2018], p. 11.
11 Cf. RICS [2014], p. 49.
12 Refers to Valuers certification (e.g. Member of the Royal Institution of Surveyors MRICS, or Member of the Appraisal Institute MAI). The review may also check if the valuer act as an independent/objective valuation expert or consultant.

13 Cf. Hackelberg [2018], p. 367.
14 Cf. Deutsche Börse [2012], p. 8.
15 Cf. Peemöller [2012], p. 242.
16 Cf. Hackelberg [2018], p. 367.
17 Simon [2006], p. 6.
18 Cf. Hackelberg [2018], p. 368.
19 Cf. Simon [2006], p. 6.
20 Cf. Hackelberg [2018], p. 368.
21 Cf. Hackelberg [2018], p. 369.
22 The checklist allows for the structured review and a first qualitative assessment of an expert opinion. However, due to the individuality of each review and valuation report this checklist cannot claim to be exhaustive; Cf. Hackelberg [2018], p. 369.

Literature and Sources

Adam, Brigitte [2018], Wo bleibt der Sachverstand? *Immobilienmanager.* Retrieved online: www.immobilienmanager.de/wo-bleibt-der-sachverstand-11042018

Appraisal Institute [2004], *The Appraisal of Real Estate,* 12 Edition, Appraisal Institute Illinois, Chicago (IL).

ASA [2011], *Valuing Machinery and Equipment: The Fundamentals of Appraising Machinery and Technical Assets, 3.* Edition, American Society of Appraisers, Reston (VA).

ASA [2021], *Valuing Machinery and Eq: The Fundamentals of Appraising Machinery and Technical Assets, 4.* Edition, American Society of Appraisers, Reston (VA).

Australian Bankers Association [2019], Industry Guideline: Appointing property valuers when lending to small businesses and primary producers, May 2019, Retrieved online: www.ausbanking.org.au/wp-content/uploads/2019/05/ABA-132351-v1-Industry_guideline_-_Appointing_property_valuers.pdf

Australian Government [2022], Market valuation for tax purposes, October 2022, Retrieved online: www.ato.gov.au/law/view/pdf/adhoc-sgml/market-valuation-guide.pdf

Baum, Andrew, Mackmin, David, and Nunnington, Nick [2018], *The Income Approach to Property Valuation,* 7. Edition, Routledge, London.

Baumann, Clayton and Lopatnikov, Alexander [2017], Scaling Laws: Uses and Misuses in the Industrial Plant and Equipment Replacement Cost Estimates, *The MTS Journal,* vol. 33, no. 2, 2nd Quarter/2017, pp. 38–44.

Becker, Klaus, Castedello, Marc, and Esser, Ingeborg [2014], *Wirtschaftsprüfung, Rechnungslegung,* Beratung, Band III, Institut der Wirtschaftsprüfer in Deutschland e.V., IDW Verlag, Dusseldorf.

Brauer, Kerry [2019], *Grundlagen der Immobilienwirtschaft Recht, Steuer, Marketing, Finanzierung, Bestandsmanagement, Projektentwicklung,* 10. Edition, Springer Gabler, Germany.

Brealey, Richard and Myers, Stewart [2000], *Principles of Corporate Finance,* 6. Edition, Irwin McGraw-Hill, Boston (MA).

Brealey, Richard and Myers, Stewart [2011], *Principles of Corporate Finance,*10. Edition, Irwin McGraw-Hill, Boston (MA).

BVS [2024], Bewertung von beweglichen Sachanlagen, in BVS-Standpunkt 6/2024, Bundesverband öffentlich bestellter und vereidigter sowie qualifizierter Sachverständiger e. *V. Fachausschuss für Maschinen, Anlagen und Betriebseinrichtungen sowie Unternehmensbewertung und Betriebswirtschaft,* Berlin, Germany.

Chinese Valuation Standards [2013], *Chinese Valuation Standards 2013,* China Appraisal Society, Wolters Kluwer CCH, Hong Kong.

Corgel, John B., Ling, David C, and Smith, Halbert C. [2001], *Real Estate Perspectives An Introduction to Real Estate,* 4. Edition, McGraw Hill Higher Education, New York.

Cotts, David G. [1997], *The Facility Management Handbook,* 2. Edition, Amacom.

Crane, Matthew [2018], The Legend of WARA and Benchmarking Purchase Price Allocation Data, November 2018, Retrieved online: https://digitalcommons.sacredheart.edu/cgi/viewcontent.cgi?article=1011&context=wcob_theses#:~:text=Introduction,of%20reasonableness%20under%20audit%20standards

Damodaran, Aswath [2005a], Misconceptions about Valuations, Retrieved from: https://pages.stern.nyu.edu/~adamodar/podcasts/valfall16/valsession2.pdf

Damodaran, Aswath [2025b], The Dark Side of Valuation: Bias, Uncertainty and Complexity, Retrieved from: https://pages.stern.nyu.edu/~adamodar/pdfiles/country/ValuationBermudaTriangle.pdf, New York.

Deutsche Börse, CMS [2012], Valuation of Chinese Companies. The Perspective of the Private Equity Industry, Deutsche Börse AG und CMS Hasche Sigle, Frankfurt am Main, Germany.

Dysert, Larry [2003], Sharpen Your Cost Estimating Skills, *Cost Engineering*, vol. 45, no. 6, pp. 22–30.

Edge, John A. [2001], The Globalization of Real Estate Appraisal: A European Perspective, *The Appraisal Journal*, vol. 69, no. 1, pp. 84–94.

Ellsworth, Richard [2007], Cost to Capacity Factor Development for Facility Projects, *Cost Engineering*, vol. 49, no. 9, p. 27.

Ellsworth, Richard [2009], Capacity Factor Cost Modeling for Gas-Fired Power Plants, *Construction Accounting & Taxation*, vol. 19, no. 1, p. 31.

EY [2022], *IFRS versus German GAAP Overview of the Main Differences*, Ernst & Young GmbH Wirtschaftsprüfungsgesellschaft, Munich, Germany.

Fahrländer, Stefan and Kloess, Stephan [2024a], *Grundlagen der Immobilienökonomie*, 2. Edition, Schweizer Immobilienschätzer Verband, St. Gallen, Switzerland.

Fahrländer, Stefan and Kloess, Stephan [2024b], *Angewandte Immobilienökonomie*, 2. Edition, Schweizer Immobilienschätzer Verband, St. Gallen, Switzerland..

Fierz, Kaspar [2001], *Wert und Zins bei Immobilien, Die neue Lehre der Immobilienbewertung und ihre Anwendung in der Praxis, 4. überarbeitete und erweiterte Auflage*, erschienen in der Schriftenreihe der Treuhand-Kammer Band 56, Zürich, Switzerland.

Floyd, Charles F. and Allen, Marcus T. [2002], *Real Estate Principles* 7. Edition, Dearborn, Chicago (IL).

Gärtner, Sven [1996], *Beurteilung und Bewertung alternativer Planungsentscheidungen im Immobilienbereich mit Hilfe eines Kennzahlensystems*, VWF Verlag für Wissenschaft und Forschung, Berlin, Germany.

Geltner, David and De Neufville, Richard [2018], *Flexibility and Real Estate Valuation under Uncertainty – A Practical Guide for the Developer*, Wiley Blackwell, Hoboken (NJ).

Geppert, Hubert and Werling, Ullrich (eds.) [2011], *Wertermittlung von Immobilieninvestments*, Manager Verlag, Cologne, Germany.

Gondring, Hanspeter [2009], *Immobilienwirtschaft – Handbuch für Studium und Praxis*, 2. Edition, Verlag Franz Vahlen, Munich, Germany.

Gondring, Hanspeter [2013], *Immobilienwirtschaft. Handbuch für Studium und Praxis*, 4. Edition, Vahlen, Munich, Germany.

Grasskamp, James A. [1972], A Rational Approach to Feasibility Analysis, *The Appraisal Journal*, vol. 40, pp. 513–521.

Hackelberg, Florian and Chan, Nova [2021], Real Estate Valuation in China, in: Wang, Bing and Just,Tobias (Eds.), *Understanding the Chinese Real Estate Market*, pp. 100–115, Springer, Cham, Switzerland.

Hackelberg, Florian and Hennig, Dirk [2021], *Investing in German Real Estate, A Practical Guide*, 1. Edition, Haufe Freiburg, Germany.

Hackelberg, Florian [2010], *Immobilienbewertung in China – unter besonderer Berücksichtigung der wirtschaftlichen, regulatorischen und materiellen Einflussfaktoren*, Wiesbaden IZ Verlag Edition, Wiesbaden, Germany.

Hackelberg, Florian [2018], Managing compliance valuation risks in M&A, in China in: Senff, Philipp and Zhang, Richard (Eds.), *Governance, Risk and Compliance Management in China*, 1. Edition, pp. 333–346, Haufe, Freiburg, Germany.

Hagen, Louise and Reif, Achim [2004], *Europäische Immobilienmärkte und ihre Bewertungsverfahren*, Verband deutscher Pfandbriefbanken, Berlin, Germany.

Hamer, Eberhard and Gebhard, Reiner [1996], *Grund- und Immobilieneigentum in der Marktwirtschaft*, Ullstein, Berlin, Germany.

Hausmann, Andrea, Rolf, Andrea, and Kröll, Ralf [2015], *Rechte und Belastungen in der Immobilienbewertung*, 5. Edition, Werner Verlag, Munich, Germany.

Hellerforth, Michaele [2012], *BWL für die Immobilienwirtschaft* 2. Edition Oldenbourg Verlag, Munich, Germany.

Hersberger, David [2008], *Wertermittlung mit dem DCF-Verfahren bei Wohn- und Büroimmobilien in der Schweiz und Deutschland,* 1. Auflage IZ Immobilien Zeitung Verlagsgesellschaft, Wiesbaden, Germany.

HKIS [2017], Hong Kong Institute of Surveyors, HKIS Valuation Standards 2017, Retrieved online: www.hkis.org.hk/ufiles/gpd-hkvs201707.pdf

Huber, Raphael Tobias [2019], Rechtsgrundlagen der Immobilienwirstchaft: Mietrecht, in: Brauer, Kerry (Ed.), *Grundlagen der Immobilienwirtschaft Recht, Steuer, Marketing, Finanzierung, Bestandsmanagement, Projektentwicklung,* 10. Edition, Springer Gabler, Germany.

Humphreys, Kenneth K. [1991], *Jelen's Cost and Optimization Engineering,* pp. 382–383, 386. McGraw-Hill, Inc., New York, NY

Iannacito, Alan C. [1989], Identification of Machinery and Equipment, in: American Society of Appraisers (Ed.), *Appraising Machinery and Equipment,* American Society of Appraisers, Washington, DC.

IAS 40 [2000], Investment Property, International Accounting Standards Board.

IDW S10 [2013], *Grundsätze der Bewertung von Immobilien,* Institute der Wirtschaftsprüfer, Standards 10. Duesseldorf, Germany.

IFRS [2018], Conceptual Framework for Financial Reporting, by the International Accounting Standards Board in September 2010. It was revised in March 2018.

IFRS 13 [2011], Fair Value Measurement, International Accounting Standards Board.

IFRS 16 [2025], Property, Plant and Equipment, International Accounting Standards Board.

IfS [1999], Leitsätze des IfS (Institut für Sachverständigenwesen), 2. Edition.

IHK [2023], IHK-München Glossar Wert- und Kostenbegriffe im Sachverständigenwesen, Retreived online: www.ihk-muenchen.de/ratgeber/sachverstaendige/wertbegriffe-und-kostenbegriffe/.

Isaac, David and O'Leary, John [2013], *Property Valuation Techniques,* 3. Edition, Palgrave Macmillan, New York.

Isenhöfer, Björn and Väth, Arno [2000], Immobilienanalyse, in: Schulte, Karl-Werner, Bone-Winkel, Stephan, and Schäfers, Wolfgang (Eds.), *Immobilien-Ökonomie 1, Betriebswirtschaftliche Grundlagen,* 2. Edition, pp. 363–424, De Gruyper Oldenburg, Berlin, Germany.

IVSC [2003], *International Valuation Standards (IVS),* 6. Edition, International Valuation Standards Committee (IVSC), London International Valuation Standards Committee.

IVSC [2005], *International Valuation Standards (IVS),* Guidance Note 3, Valuation of Plant and Equipment (Revised 2005), effective as of 01.01.2026.

IVSC [2025], *International Valuation Standards (IVS), International Valuation Standards Committee (IVSC)*, London, International Valuation Standards Committee.

Junius, Karsten and Piazolo, Daniel [2008], Research in der Grundstücksbewertung, in: Junius, K., and Piazolo, D. (Eds.), *Praxishandbuch Immobilien-Research*, pp. 147–170, Immobilien Manager Verlag, Cologne, Germany.

Kleiber, Wolfgang and Simon, Reinholt [2007], *Verkehrswertermittlung von Grundstücken, Kommentar und Handbuch*, 5. Edition, pp. 147–170, Bundesanzeiger Verlag, Cologne, Germany.

Kleiber, Wolfgang [2017], *Verkehrswertermittlung von Grundstücken, Kommentar und Handbuch,* 8. Edition, Bundesanzeiger Verlag, Cologne, Germany.

Kleiber, Wolfgang [2018], *Wertermittlung nach ImmoWertV. Praxiskommentar zur Verkehrswertermittlung von Grundstücken*, 8. Edition, Bundesanzeiger Verlag, Cologne, Germany.

Kleiber, Wolfgang [2020], Marktwertermittlung nach ImmoWertV, Kommentar und Handbuch zur Ermittlung von Marktwerten (Verkehrswerten) und Beleihungswerten sowie zur steuerlichen Bewertung unter Berücksichtigung der ImmoWertV, *Bundesanzeigerverlag*, 9. Edition, Cologne, Germany.

Koller, Tim, Goedhardt, Marc, and Wessels, David [2010], *Valuation – Measuring and Managing, The Value of a Company*, 5. Edition, Wiley, Hoboken (NJ).

KPMG [2024], *Fair Value Measurement Handbook US GAAP and IFRS Accounting Standards*, KPMG UK.

Kruschwitz, Lutz, Löffler, Andreas, and Essler, Wolfgang [2009], *Unternehmensbewertung für die Praxis – Fragen und Antworten*, Schaeffer Poeschel, Stuttgart, Germany.

Leopoldsberger, Gerrit, Thomas, Matthias, and Naubereit, Philipp [2016], Immobilienbewertung, in: Schulte, Karl-Werner, Bone-Winkel, Stephan, and Schäfers, Wolfgang, *Immobilien-Ökonomie 1, Betriebswirtschaftliche Grundlagen*. 5. Edition, pp. 425–480, De Gruyper Oldenburg, Berlin, Germany.

Leopoldsberger, Gerrit [2017], Wertermittlung von Immobilien, in: Mändle, Markus (Ed.), *Handbuch Immobilienwirtschaft*, Haufe, Freiburg, Germany, pp. 479–518.

Manganelli, Benedetto [2015], *Real Estate Investing – Market Analysis, Valuation Techniques, and Risk Management*, Springer, Heidelberg, Germany.

Matysiak, Georg A. [2017], Automated Valuation Models (AVMs): A Brave New World? Conference: Wroclaw Conference in Finance, Retrieved from: www.researchgate.net/publication/319355261_Automated_Valuation_Models_AVMs_A_brave_new_world

Metzger, Bernhard [2018], *Wertermittlung von Immobilien und Grundstücken*, 6. Edition, Haufe Freiburg, Germany.

Meyer, Jens [2007], *Europäische Immobilienbewertungsstandards, Methoden und Begriffe*, VDM Verlag, Dr. Müller, Berlin, Germany.

Miller, George A. [1956], The Magical Number Seven, Plus or Minus Two: Some Limits on Our Capacity for Processing Information, *Psychological Review*, vol. 63, pp. 81–97.

NIESV [2015], The Nigerian Valuation Standards ("Green Book"), The Nigerian Institute of Estate Surveyors and Valuers (Faculty of Valuation and Compensation), Retrieved online: https://niesv.org.ng/newsdocs/Green_Book.pdf

O'Mara, Martha A. [1999], *Strategy and Place – Managing Corporate Real Estate and Facilities for Competitive Advantage*, Simon and Schuster Inc, New York.

Parker, David [2016], *International Valuation Standards, A Guide to the Valuation of Real Property Assets*, Wiley Blackwell, Hoboken (NJ).

Peemöller, Volker H. [2012], *Praxishandbuch der Unternehmensbewertung Grundlagen und Methoden, Bewertungsverfahren, Besonderheiten bei der Bewertung*, 5. Edition, Verlag Neue Wirtschaftsbriefe, Berlin, Germany.

Petersen, Hauke [2005], *Verkehrswertermittlung von Immobilien – Praxisorientierte Bewertung*, Richard Boorberg Verlag, Kiel, Germany.

Petersen, Hauke [2008], *Marktorientierte Immobilienbewertung – Grundriss für die Praxis*, 8. Edition, Richard Boorberg Verlag, Kiel, Germany.

Petersen, Hauke, Schnoor, Jürgen, and Seitz, Wolfgang [2018], *Verkehrswertermittlung von Immobilien, Praxisorientierte Bewertung*, 3. Edition, Boorberg, Munich, Germany.

Petersen, Hauke, Schnoor, Jürgen, and Seitz, Wolfgang [2019], *Marktorientierte Immobilienbewertung*, 10. Edition, Boorberg, Munich, Germany.

Pfnür, Andreas [2011], *Modernes Immobilienmanagement, Immobilieninvestment, Immobiliennutzung, Immobilienentwicklung und -betrieb*, 3. Edition, Springer Heidelberg, Dorbrecht, London, New York.

Pyhrr, Stephen A. [1989], *Real Estate Investment – Strategy, Analysis, Decision*, 2. Edition, John Wiley & Sons Inc, New York.

Plock, Michael, Kadel, Dirk, and Hennig, Dirk, [2025], Jahrestagung – PwC Workshop Bewertung von Maschinen/und Anlagen für Finanzierungszwecke am Praxisbeispiel, 27 March 2025, Berlin, Germany.

Preuß, Norbert and Schöne, Lars Bernhard [2016], *Real Estate Facility Management, Aus Sicht der Consulting Praxis*, 4. Edition, Springer Vieweg, Berlin, Germany.

Reinert, Jan [2021], *Property Valuations and Valuation Accuracy in Germany and Europe*, Dissertation zur Erlangung des Grades eines Doktors der Wirtschaftswissenschaft eingereicht an der Fakultät für Wirtschaftswissenschaften der Universität Regensburg, Regensburg.

Reisbeck, Tillmann and Söhne, Lars Bernhard [2009], *Immobilien Benchmarking, Ziele, Nutzen, Methoden und Praxis*, 2. Edition, Springer, Berlin, Germany.

Renner, Ulrich and Sohni, Michael [2012], *Ross-Bachmann, Ermittlung des Verkehrswertes von Immobilien,* 3. Edition, Theodor Oppermann Verlag, Hanover, Germany.

RICS [2007a], *RICS Valuation Standards ("The Red Book",* 6. Edition, The Royal Institution of Chartered Surveyors (RICS), London.

RICS [2008], *Die Wertermittlungsstandards der RICS German* 6. Edition, The Royal Institution of Chartered Surveyors (RICS), Coventry, UK/Frankfurt a. M.

RICS [2013], *Automated Valuation Models (AVMs) RICS Information Paper,* 1. Edition, Royal Institution of Chartered Surveyors, London.

RICS [2014], *RICS Valuation – Global Standards, Royal Institution of Chartered Surveyors (RICS),* London.

RICS [2017], *The Future of Valuations The Relevance of Real Estate Valuations for Institutional Investors and Banks – Views from a European Expert Group,* Royal Institution of Chartered Surveyors, London.

RICS [2022], *RICS Valuation – Global Standards,* effective from 31 January 2022, Royal Institution of Chartered Surveyors, London.

RICS [2024], *RICS Valuation – Global Standards Basis for Conclusion*, December 2024, effective from 31 January 2025, Royal Institution of Chartered Surveyors, London. Retrieved online; www.rics.org/content/dam/ricsglobal/documents/standards/Red-Book-2024_Basis-for-Conclusions.pdf

Salden, Georg [2014], *Die dynamische Methode, Erfolgreich in Immobilien investieren durch perfektes Timing und ganzheitliche Betrachtung,* Haufe, München, Freiburg, Germany.

Schulte, Karl-Werner, Bone-Winkel, Stephan, and Schäfers, Wolfgang [2016], *Immobilien-Ökonomie 1, Betriebswirtschaftliche Grundlagen,* 5. Edition, De Gruyper Oldenburg, Berlin, Germany.

Sehab, Hatem [2025], The Science behind Approach, Methodology & Framework – Are They Really Different? Retrieved from: www.linkedin.com/pulse/science-behind-approach-methodology-framework-really-different-hatem/

Simon, Thore [2006], *Plausibilisierung von Verkehrswerten (Marktwerten), Ellwood-Verfahren und prognostische DCF-Modelle*, Deutsche Immobilien-Akademie an der Universität Freiburg GmbH, Freiburg, Germany.

Simon, Thore and Gillich, Tobias [2012], *Wertermittlung von Grundstücken*, 6. Edition, Werner Verlag, Munich, Germany.

Sommer, Götz and Kröll, Ralf [2017], *Lehrbuch zur Immobilienbewertung,* 5. Edition, Werner Verlag, Munich, Germany.

Stand, Moritz [2023], *Real Estate Valuation in the Age of Artificial Intelligence - Modern Machine Learning Algorithms and their Application in Property Appraisal*. Submitted to the Faculty of Business, Economics, and Management Information Systems at the University of Regensburg, Regensburg, Germany.

TEGOVA [2003], *European Valuation Standards,* 5. Edition, The European Group of Valuers' Association.

TEGOVA [2016], *European Valuation Standards (EVS 6) Automated Valuation Models (AVMs)*, The European Group of Valuers' Association.

TEGOVA [2022], *European Plant, Machinery and Equipment Valuation Standards,* 1. Edition, The European Group of Valuers' Association.

TEGOVA [2025], *European Valuation Standards,* 10. Edition, The European Group of Valuers' Association.

Thomas, Matthias [2011], Immobilienbewertung, in: Rottke, Nico and Thomas, Matthias (Eds.), *Immobilienwirtschaftslehre, Band 1 Management,* Manager Verlag, Cologne, Germany, pp. 763–834.

Thöne, Christian [2006], *Grundlagen der internationalen Immobilienbewertung, Skript einer Vorlesung an der Bauakademie*, Biberach, Leipzig, Germany.

Ulaganathan, Mani and Sugumar, S. [2024], Engineering Economics – Real Estate Asset Valuation – Study of Economics – Cost (to) the Capacity Method, *Educational Administration Theory and Practice*, vol. 30, no. 4, pp. 9741–9749.

Vogel, Roland R. [2000], Angelsächsische Investitionsverfahren und marktorientierte Verkehrswertermittlung in Deutschaland, erschienen in *Grundstücksmarkt und Grundstückswert* (GuG), vol. 4/2020, S. 210 – 213, Köln, Germany.

White, Darron, Turner, John, Kenyon, Bruce, and Lincoln, Nicole [2003], *Internationale Bewertungsverfahren für das Investment in Immobilien*, Verlag Immobilienzeitung, 3. Edition, Aktualisierte Auflage, Wiesbaden, Germany.

Wöhle, Claudia B. [2005], Moderne Immobilienbewertung mit Hilfe der Discounted Cashflow-Verfahren, in: *Immobilienmärkte und Immobilienbewertung*, Franke, Hans-Hermann, Rehkugel, Heinz, Vahlen Verlag, München, Germany.

Wyatt, Peter [2007], *Property Valuation in an Economic Context*, Blackwell Publishing, Bristol, England.

Wyatt, Peter [2013], *Property Valuation*, 2. Edition, Wiley Blackwell, Hoboken (NJ).

Zimmermann, Peter [2010], *ImmoWertV – Immobilienwertermittlung*, Kommentar, C.H.Beck, Munich, Germany.

Zyla, Mark L. [2013], *Fair Value Measurement: Practical Guidance and Implementation,* Wiley, Hoboken (NJ).

Index

Note: Page numbers in *italics* refer to figures.

For Product Safety Concerns and Information please contact our EU
representative GPSR@taylorandfrancis.com
Taylor & Francis Verlag GmbH, Kaufingerstraße 24, 80331 München, Germany

www.ingramcontent.com/pod-product-compliance
Ingram Content Group UK Ltd.
Pitfield, Milton Keynes, MK11 3LW, UK
UKHW052212190926
475702UK00006B/975

* 9 7 8 1 0 4 1 0 1 7 4 3 1 *